# PROGRESS IN CHEMICAL PHYSICS RESEARCH

# PROGRESS IN CHEMICAL PHYSICS RESEARCH

**A. N. LINKE**
**EDITOR**

**Nova Science Publishers, Inc.**
*New York*

For permission to use material from this book please contact us:
Telephone 631-231-7269; Fax 631-231-8175
Web Site: http://www.novapublishers.com

**NOTICE TO THE READER**



LIBRARY OF CONGRESS CATALOGING-IN-PUBLICATION DATA
Progress in chemical physics research / Linke, A.N. (editor).
p. cm.
Includes bibliographical references and index.
ISBN 1-59454-412-3 -- ISBN 1-59454-451-4
1. Chemistry, Physical and theoretical. I. Linke, A. N.
QD453.3.P76 2005
541--dc22 2005010779

*Published by Nova Science Publishers, Inc. ✦ New York*

# CONTENTS

# PREFACE

Chemical physics and physical chemistry are closely related fields of study. Together they are distinguished from other disciplines by the incredible range of problems addressed by their practitioners. An effective physical chemist or chemical physicist is a "jack-of-all-trades," able to apply the principles and techniques of the field to everything from high-tech materials to biology. Just as the fields of chemistry and physics have expanded, so have chemical physics subject areas, which include polymers, materials, surfaces/interfaces, and biological macromolecules, along with the traditional small molecule and condensed phase systems. This new book gathers research from around the world presenting important new developments.

In Chapter 1, the authors present an emerging fields of quantum chemistry at finite temperature. They discuss its recent developments on both theoretical and experimental fronts. They describe and analyze several experimental investigations related to the temperature effects on the structure, electronic spectra, or bond rupture forces for molecules. This includes the study of the temperature impact on the pathway shifts for the protein unfolding by atomic force microscopy (AFM), the temperature dependence of the absorption spectra of electrons in solvents, and temperature influence over the intermolecular forces measured by the AFM. On the theoretical side, we review a recent advancement made by the author in the coming fields of quantum chemistry at finite temperature. Starting from Bloch equation, we have derived the sets of hierarchy equations for the reduced density operators in both canonical and grand canonical ensembles. They provide a law according to which the reduced density operators vary in temperature for the identical and interacting many-body particles. By taking the independent particle approximation, we have solved the equation in the case of a grand canonical ensemble, and obtained an eigenequation for the molecular orbitals at finite temperature. The explicit expression for the temperature-dependent Fock operator is also given. They will form a foundation for the study of the molecular electronic structures and their interplay with the finite temperature. Furthermore, we clarify the physics concerning the temperature effect on the electronic structure or processes of the molecules which is crucial for both theoretical understanding and computational study. Finally, we summarize our discussion and point out the theoretical and computational issues for the future explorations in the fields of quantum chemistry at finite temperature.

The phenomenology of phase transitions in free atomic and molecular clusters is reviewed in Chapter 2, based on how it is revealed from numerical simulations and experiments. The authors mainly consider transitions involving condensed phases, paying a particular attention at the finite size analog of melting as well as solid-solid transitions. The chosen examples range from weakly bound van der Waals clusters to ionic or metal clusters. In general they attempt to correlate the observed behavior to the features of the energy landscape, and possible indicators to characterize phase changes are critically discussed. The transition to the vapor phase is mentioned with a view at evaporation phenomena, and a way to quantify the cluster boiling point is proposed.

Recently the first examples of chemical systems which show the amplification of enantiomeric excess (ee) were found. Inspired by these experiments, the authors have previously proposed in Chapter 3 a chemical reaction model which leads to chirality selection in a closed system. Here they extend the model and analyze general ones in detail by means of flow diagrams in the phase space of enantiomer concentrations. With only a quadratic autocatalytic process in the production of chiral molecules from achiral substrate, ee is found to be amplified, but the flow is attracted to a fixed line and the final degree of ee depends strongly on the initial condition. When the produced chiral molecules decompose back to achiral substrate, they are recycled in the catalytic production, because the system is closed. Various recycling processes are found to alter the final state of the system completely: With a cross inhibition and a linear autocatalysis a chiral symmetry of the system is broken and a complete homochirality is achieved. With a back reaction only, the quadratic autocatalysis is found necessary to establish chirality selection.

Chapter 4 reviews previous work on the dynamic structure factor $S(\kappa,\omega)$ of liquid Ge ($\ell$-Ge) at temperature T = 1250 K, and of amorphous Ge (a-Ge) at T = 300 K, using *ab initio* molecular dynamics. The electronic energy is computed using density-functional theory, primarily in the generalized gradient approximation, together with a plane wave representation of the wave functions and ultra-soft pseudopotentials. The authors use a 64-atom cell with periodic boundary conditions, and calculate averages over runs of up to about 16 ps. The calculated liquid $S(\kappa,\omega)$ agrees qualitatively with that obtained by Hosokawa *et al*, using inelastic X-ray scattering. In a-Ge, they find that the calculated $S(\kappa,\omega)$ is in qualitative agreement with that obtained experimentally by Maley *et al.* The results suggest that the *ab initio* approach is sufficient to allow approximate calculations of $S(\kappa,\omega)$ in both liquid and amorphous materials.

It is shown in Chapter 5 that the dependence of negative thermal expansion coefficient of $ZrW_2O_8$ and $HfW_2O_8$ on the temperature found by many authors contradicts to an important thermodynamic relation $(\partial C_P/\partial P)_T = - TV((\partial\alpha/\partial T)_P + \alpha^2)$.

It is shown in Chapter 6 that the dependence of negative thermal expansion coefficient of many substances on the temperature contradicts to an important thermodynamic relation $(\partial C_P/\partial P)_T = - TV((\partial\alpha/\partial T)_P + \alpha^2)$. It is supposed that there are oscillations at the $C_P(T)$ and $\alpha(T)$ curves at $\alpha > 0$ for some substances and they are in reverse phases.

Crystal orientation effects on electronic and optical properties of strained GaN semiconductors are reviewed in Chapter 8. Analytical expressions for the band gap, wave function, and momentum matrix element of bulk GaN semiconductors with an arbitrary crystal orientation are shown using a block-diagonalized Hamiltonian. Self-consistent (SC) solutions considering the spontaneous (SP) and piezoelectric (PZ) polarizations for valence-band structures of a quantum well (QW) with an arbitrary crystal orientation are obtained by solving the Schrödinger equation for electrons, the $6 \times 6$ Hamiltonian for holes, and Poisson's equation iteratively. The optical gain is investigated using the non-Markovian gain model with many-body effects. The optical gain increases with increasing crystal angle, which is mainly due to the increase of the matrix elements with increasing crystal angle. In particular, the optical gain of the (10$\bar{1}$0)-oriented QW structure is largely enhanced over that of the (0001)-oriented QW structure. This can be explained by the fact that the optical matrix elements near the band edge (k=0) of the (10$\bar{1}$10)-oriented QW structure significantly increase compared to those of the (0001)-oriented structure, in addition to the reduction of the effective mass. The threshold current density Jth of the (10$\bar{1}$10)-oriented QW structure is expected to be about a half of that of the (0001)-oriented QW structure.

In: Progress in Chemical Physics Research
Editor: A. N. Linke, pp. 1-20

ISBN 1-59454-451-4


*Chapter 1*

# ON EMERGING FIELDS OF QUANTUM CHEMISTRY AT FINITE TEMPERATURE

***Liqiang Wei***
Institute for Theoretical Atomic, Molecular and Optical Physics
Harvard University, Cambridge, MA 02138

**Abstract**

In this article, we present an emerging fields of quantum chemistry at finite temperature. We discuss its recent developments on both theoretical and experimental fronts. We describe and analyze several experimental investigations related to the temperature effects on the structure, electronic spectra, or bond rupture forces for molecules. This includes the study of the temperature impact on the pathway shifts for the protein unfolding by atomic force microscopy ($AFM$), the temperature dependence of the absorption spectra of electrons in solvents, and temperature influence over the intermolecular forces measured by the $AFM$. On the theoretical side, we review a recent advancement made by the author in the coming fields of quantum chemistry at finite temperature. Starting from *Bloch* equation, we have derived the sets of hierarchy equations for the reduced density operators in both canonical and grand canonical ensembles. They provide a law according to which the reduced density operators vary in temperature for the identical and interacting many-body particles. By taking the independent particle approximation, we have solved the equation in the case of a grand canonical ensemble, and obtained an eigenequation for the molecular orbitals at finite temperature. The explicit expression for the temperature-dependent *Fock* operator is also given. They will form a foundation for the study of the molecular electronic structures and their interplay with the finite temperature. Furthermore, we clarify the physics concerning the temperature effect on the electronic structure or processes of the molecules which is crucial for both theoretical understanding and computational study. Finally, we summarize our discussion and point out the theoretical and computational issues for the future explorations in the fields of quantum chemistry at finite temperature.

# 1 Introduction

The history for quantum chemistry development is almost synchronous to that of quantum mechanics itself. It begins with *Heitler* and *London*'s study of electronic structure of $H_2$ molecule shortly after the establishment of wave mechanics for quantum particles [1]. There are two major types of molecular electronic theories: valence bond approach *vs.* molecular orbital method with the latter being the popular one for the present investigation. It has gone through the stages from the evaluation of molecular integrals via a semiempirical way to the one by an *ab initio* method. Correlation issue is always a bottleneck for the computational quantum chemistry and is under intensive study for over fifty years. For large molecular systems such as biomolecules and molecular materials, the development of the combined $QM/MM$ approach, pseudopotential method and linear scaling algorithm has significantly advanced our understanding of their structure and dynamics. There are about eight Nobel prize laureates whose researches are related to the molecular electronic structure theory. This not only recognizes the most eminent scientists who have made the outstanding contributions to the fields of quantum chemistry, but more importantly, it indicates the essential roles the electronic structure theory has been playing in the theoretical chemistry as well as for the whole areas of molecular sciences. Nowadays, quantum chemistry has been becoming a maturing science [2, 3].

Nevertheless, the current fields of quantum chemistry are only part of the story for the molecular electronic structure theory. From the pedagogical points of view, the quantum mechanics based on which the traditional quantum chemistry is built is a special case of more general quantum statistical mechanics [4, 5, 6]. In reality, the experimental observations are performed under the conditions with thermodynamic constraints. Henceforth, there is a need to extend the current areas of quantum chemistry to the realm of, for instance, finite temperature [4, 5, 6].

Indeed, many experimental investigations for various fields and for different systems have already shown the temperature or pressure effects on their microscopic structures [7-30,49-57,63,71-87]. The polymeric molecules are one of the most interesting systems for this sort of studies [7-16]. The experimental measurement on the absorption spectra, photoluminescence ($PL$), and photoluminescence excitation ($PLE$), and spectral line narrowing ($SLN$) for the $PPV$ and its derivatives all show the same trend of the blue shift with the increasing temperature [7, 8, 9]. This attributes to the temperature dependence of their very rich intrinsic structures such as the vibronic coupling [14, 15, 16]. The experimental investigation of the temperature effect on the biomolecules started in the late nineteenth century [17, 18]. Most recently, it has been extended to the study of folding and unfolding of protein or $DNA$ [19, 20, 21]. In addition to the observed patterns for the unfolding forces with respect to the extension or temperature, it has been proved that the temperature-induced unfolding is another way for the study of mechanisms or pathways of protein folding or unfolding processes [19-23]. The newest related development is on the $AFM$ measurement made by *Lo* et al. of the intermolecular forces for the biotin-avidin system in the temperature range from 286 to $310K$ [63]. It has shown that an increase of temperature will almost linearly decrease the strength of the bond rupture force for the individual biotin-avidin pair. The study of temperature effect on the absorption spectra of solvated electron began in the 1950's and it is still of current interest. A striking effect is that an increasing temperature will cause the positions of their maximal absorption red shift [71-85].

In recent papers, we have deduced an eigenequation for the molecular orbitals [4, 5]. It is the extension from the usual $Hartree - Fock$ equation at zero temperature to the one at any finite temperature [88, 89]. It opens an avenue for the study of the temperature effects on the electronic structures as well as their interplay with the thermodynamic properties. In the third section, we will present this equation and give the details for its derivation. In the next section, we will show four major types of experiments related to the study of the temperature influences over the microscopic structure of molecular systems. In the final section, we will discuss and analyze our presentations, and point out both theoretical and computational issues for the future investigation.

# 2 Experimental Development

In this section, we mainly describe the experimental investigations related to the temperature effect on the bonding, structure and electronic spectra of molecules. We choose four kinds of the most recent developments in these fields which are of chemical or biomolecular interests.

## 2.1 Temperature Effects on Geometric Structure and UV-Visible Electronic Spectra of Polymers

The first important systems where the important issues related to the temperature effect on the geometric structure and electronic spectra are the polymeric molecules. Many experimental investigations and some theoretical work already exist in the literature [7-16]. However, how the temperature changes the microscopic structures of the polymers are still not completely understood and there are many unresolved issues in interpreting their electronic spectra. We list here a few very interesting experimental investigations for the purpose of demonstration.

The poly($p$-phenylenevinylene)($PPV$) is one of the prototype polymeric systems for the study of their various mechanical, electronic, and optical properties. The impact from the temperature on the absorption spectra, photoluminescence ($PL$), and photoluminescence excitation ($PLE$) of the $PPV$ have also been investigated both experimentally and theoretically [7, 8, 9]. In an experiment carried out by $Yu$ et al., the absorption spectra are measured for the $PPV$ sample from the temperature 10 to $330K$. The details of the experiment are given in their paper [9]. The resulting spectra for the absorption at $T = 80$ and $300K$ are shown in Figure 1 of that paper. We see that there is a pronounced change in the spectra when increasing the temperature. They also study the $PL$ and $PLE$ spectra for the $PPV$. The measured $PL$ spectra at two temperatures: 77 and $300K$ are demonstrated in the Figure 3, and the $PLE$ spectra at those temperatures are depicted in the Figure 4 of the paper [9]. They both show the dramatic changes of the band blue shift when the temperature is increased. Similar studies have also been performed before by the other groups [7, 8]. They observed the similar behaviors.

Another interesting investigation is related to the temperature effect on the spectral line narrowing ($SLN$) of the poly(2-methoxy-5-(2$'$-ethylhexyloxy)-1,4-phenylene-vinylene) ($MEH-PPV$) spin-coated from either $THF$ or $CB$ solvents [11]. In the experiment done by Sheridan et al., the $SLN$ is measured together with the absorption and $PL$ as shown in Fig. 1 of their paper. It is found that the same trend of the $SLN$ blue shift is observed as that for the absorption and $PL$ with an increasing temperature. They attribute this to the same reason of the electronic structure modification resulting from the variation of the temperature.

## 2.2 Temperature Effects on Structure, Dynamics, and Folding/Unfolding of Biomolecules

Biomolecules are complex systems, featuring a large molecular size, a heterogeneity of atomic constitutes and a variety of conformations or configurations. Their energy landscape thereby exhibits multiple substates and multiple energy barrier, and varies in size for the barrier heights [24, 25, 26, 27, 28]. The temperature should have a strong effect on their structure and dynamics including the folding or unfolding [17-57]. This effect could be either from the fluctuation of thermal motion of molecules or due to the redistribution of electronic charge as we will discuss in the next section.

The experimental observation of the temperature impact on the microscopic structure of biological systems dates back to the very early days. One focus, for example, is on the measurement of the elastic properties of the human red blood cell membrane as a function of temperature [17, 18]. Another related study is about the influence over the thermal structural transition of the young or unfractionated red blood cells due to the involvement of protein spectrin which might modified the spectrin-membrane interaction [29, 30]. Most recently, the atomic force microscopy ($AFM$) has been used to detect the impact from the variation of the temperature on the spectrin protein unfolding force as well as on the bond rupture force for the biotin-avidin system [19, 20, 21, 63].

The $AFM$ is a surface imaging technique with an atomic-scale resolution capable of measuring *any* types of the forces as small as $10^{-18}$ $N$. It combines the principle of scanning tunnelling microscopes ($STM$) and stylus profilometer, and therefore can probe the surfaces of both conducting and nonconducting samples [31, 32]. The imaging on soft materials such as biomolecules with the $AFM$ has been performed beginning in the 1980's [33, 34, 35]. Recently, it has been applied to measure the adhesive forces and energies between the biotin and avidin pair also as we will show in the next subsection [36, 37, 38, 39]. Unlike other experimental techniques, the $AFM$ features a high precision and sensitivity to probe the surface with a molecular resolution, and can be done in physiological environments.

In an $AFM$ investigation of mechanical unfolding of titin protein, for example, the restoring forces all show a sawtooth like pattern with a definite periodicity. It reveals much information about the mechanism of unfolding processes [42, 43]. The observed pattern, in addition to fit a worm-like chain model, has also been verified by the steered molecular dynamics or Monte carlo simulations [44, 45]. Similar study has been extended to other systems [46, 47, 48].

The same kind of experiments has also been performed by varying the temperature. In the experiment carried out by Spider and Discher et al. [21], the spectrin protein is chosen for the $AFM$ study at different temperatures. Thousands of tip-to-surface contacts are performed for a given temperature because of the statistical nature of the $AFM$ measurement. The observed curve for the relation between the unfolding force and extension shows the similar sawtooth pattern for all temperatures. In addition, the tandem repeat unfolding events are more favored at lower temperature as demonstrated in the unfolding length histograms. Most striking is that the unfolding forces show a dramatically nonlinear decreasing relation as the temperature $T$ approaches the transition temperature $T_m$. This is shown in the Figure $3C$ of paper [21].

Similar behaviors regarding the force-temperature dependence have also been observed via either $AFM$ or optical tweezers for the forced overstretching transition for the individual double-stranded $DNA$ molecules [19, 20, 49, 50].

Some other interesting experiments which illustrate the effect of temperature on the microscopic structure of biomolecules have also been performed even though the detailed physical origins of the effect (from either the electrons or the molecules) have not been specified [52, 53, 54, 55, 56, 57]. In a circular dichroism ($CD$) spectra and high resolution $NMR$ study, for instance, it shows that the secondary structure of the Alzheimer $\beta$ (12-28) peptide is temperature-dependent with an extended left-handed $3_1$ helix interconverting with a flexible random coil conformation [52]. Another example is related to the analysis of the temperature-dependent interaction of the protein *Ssh10b* with $DNA$ which influences the $DNA$ topology [53, 54, 55]. The study from the heteronuclear $NMR$ and site-directed mutagenesis indicates that the *Ssh10b* exists as a dimer: $T$ form and $C$ form. Their ratio is determined by the $Leu^{61} - Pro^{62}$ peptide bond of the *Ssh10b* which is sensitive to the temperature.

## 2.3 Temperature Effects on Intermolecular Forces

The study of the general issues related to the temperature effects on the microscopic structure has been most recently extended to the realm of intermolecular forces. Since the usual intermolecular forces such as hydrogen bond, *van der Waals* force, ionic bond, and hydrophobic interaction are weak and typically of the order of $0.1\ eV$ or $4.0\ kT$ at the physiological temperature, the variation of temperature will thereby have a very strong influence over the strength of these forces.

The first experimental investigation on the temperature-dependent intermolecular forces is for the biotin-avidin system and by an $AFM$ measurement [63]. The biotin-avidin complex is a prototypical receptor and ligand system with the biotin binding strongly up to four avidin protein [58, 59, 60, 61, 62]. They have an extremely high binding affinity, and therefore serves as a model system for various experimental investigations. In the experiment carried out in *Beebe*'s group [63], the receptor avidin is attached to the $AFM$ tip and linked to the agarose bead functionalized with the biotin. The temperature of the entire $AFM$ apparatus is varied at a range from 286 to $310K$. In addition, the loading rate is kept very slow so that the thermal equilibrium for the biotin-avidin pairs is assumed. The forces expected to be determined is the rupture force $F_i$ between the individual biotin-avidin pair which is defined as the maximum restoring force [63]. In actual $AFM$ experiment, however, the total adhesive force between the tip and substrate is measured. It is a sum of finite number $n$ of the interactions between each biotin and avidin pair. To extract the individual and average bond rupture, a statistical method has been developed in *Beebe*'s group [39, 40, 64, 65, 66]. They assume a *Poisson* distribution for the number $n$ of the discrete rupture forces or linkages from multiple measurements, and have obtained the single force $F_i$ at different temperatures. The result is shown in the Figure 3 of the paper [63]. We see that the individual rupture force $F_i$ for the biotin-avidin pair is decreased by about five-fold in strength when the temperature is increased from 286 to $310K$.

To interpret the observed temperature impact on the biotin-avidin forces, *Peebe*'s group has performed a thermodynamic analysis [63]. Based on the simple models and arguments [67, 68], they have come out an equation that connects the square of the single bond-rupture force $F_i$ to the absolute temperature $T$ as follows,

$$F_i^2 = 2\Delta E^{\ddagger} k_{bond} - 2k_B T k_{bond} \ln\left(\frac{\tau_R}{\tau_D}\right) \tag{1}$$

where the $k_{bond}$ is the force constant of the individual biotin-avidin pair, and the time $\tau_R$ is the characteristic time needed to break $n$ pairs of those forces. The $E^{\ddagger}$ is the energy required to remove the biotin from avidin's strongest binding site and the corresponding time is $\tau_D$. More detailed on this analysis can be found in the paper [63]. The relation between the square of the force $F_i$ and temperature $T$ is also plotted as Figure 5 in that paper. Therefore, from the relation (1) and that Figure, the information about the stiffness of the ligand and receptor bond and the critical binding energy, etc. can be obtained. Obviously, what we need at present is a microscopic theory which can account for all these relations and properties.

## 2.4 Temperature Effects on Absorption Spectra of Electrons in Solvents

The structure and dynamics of solute in solvent is one of the most important fields in chemistry since most of the chemical reactions occur in solution phases. In the meantime, it is also one of the most challenging fields in theoretical chemistry with many unsettled issues. The variation of temperature in the measurement of absorption spectra of solvated electron in various solvents has proved to be a useful means for the understanding of the solvation processes [71-85].

There are several experimental techniques available for this type of studies with the pulse radiolysis being the most commonly used one. There are also several research groups doing the similar experimental investigations and obtaining the consistent results relating to the temperature effects on the optical absorption spectra of solvated electron in solvents. In a recent experiment carried out in $Katsumura$'s group, for example, the pulse radiolysis technique is employed to study the optical absorption spectra of the solvated electron in the ethylene glycol at different temperatures from $290$ to $598K$ at a fixed pressure of $100$ atm. In addition to the faster decay of absorptions, it is found that, their maximal positions shift to the red with the increasing temperature as shown in the Figures 1 to 3 of the paper [81]. This is in contrast to the situation for the electronic spectra of the polymers. They also point out the need to quantify the change of the density in the experiment in order to really understand the observed results.

The same type of experiment has been extended to the study of the optical absorption spectra for $Ag^0$ and $Ag_2^+$ in water by varying the temperature, and similar results have been obtained [80].

# 3 Theoretical Development

Having presented four different types of experiments above, we have seen that the temperature effect on the microscopic structure of molecules is a very interesting and sophisticated field. Many more needs to be understood and probed. Even though the experimental study has been for a long time, very limited number of the related theoretical work is available, especially at the first-principle level. In other words, the quantum chemistry at finite temperature is not a well-defined or well-established field [4, 5, 6].

It is true that the influence of temperature on the microscopic structure is a complicated phenomenon. There exists different functioning mechanisms. One consideration is that the variation of the temperature, according to the Fermi-Dirac statistics, will change the thermal probability distribution of single-particle states for a free electron gas. It is expected that similar situation should occur for an interacting electron system, and therefore its microscopic structure will be correspondingly altered. Another consideration is that, for molecules or solids, the thermal excitation will cause the change of the time scales for the molecular motions. This will most likely bring about the transitions of the electronic states, and therefore lead to the breakdown of the Born-Oppenheimer approximation. Electron-phonon interaction is a fundamental topic in solid state physics and its temperature dependence is well-known. As a result, the temperature could change the strength of the coupling between the electronic and molecular motions. Nevertheless, we tackle the issues in a simpler way. We treat only an interacting identical *fermion* system, or neglect the coupling of the electronic motion with those of the nucleus in molecules or solids. We expect that some sort of the general conclusions will come out from this study. As a matter of fact, this is also the approach usually adopted in a non-adiabatic molecular dynamics, in which purely solving the eigenequation for the electrons will provide the reference states for the investigation of the coupling motions between the electrons and the nucleus of the molecules.

In the following, we will present a self-consistent equation within the framework of equilibrium statistical mechanics which decides the molecular orbitals at a given temperature.

### 3.1 Hierarchy Bloch Equations for Reduced Density Operators in Canonical Ensemble

We consider an identical and interacting $N$-particle system. In a canonical ensemble, its $N$th-order density operator takes the form

$$D^N = \exp(-\beta H_N), \tag{2}$$

and satisfies the *Bloch* equation [90, 91]

$$-\frac{\partial}{\partial \beta} D^N = H_N D^N, \tag{3}$$

where

$$H_N = \sum_{i=1}^{N} h(i) + \sum_{i<j}^{N} g(i,j), \tag{4}$$

is the *Hamiltonian* for the $N$ particle system composed of the one-particle operator $h$ and two-body operator $g$. The $\beta$ is the inverse of the product of *Boltzmann* constant $k_B$ and absolute temperature $T$.

Since the *Hamiltonian* (4) can be written as a reduced *two*-body operator form, the second-order reduced density operator suffices to describe its $N$ $(\geq 2)$ particle quantum states. A $p$th-order reduced density operator is generally defined by [92, 93]

$$D^p = L_N^p(D^N), \tag{5}$$

where $L_N^p$ is the contraction operator acting on an $N$th-order tensor in the $N$-particle *Hilbert* space $V^N$. The trace of the $D^p$ gives the partition function,

$$Tr(D^p) = Z(\beta, V, N). \tag{6}$$

Rewrite the *Hamiltonian* in a form

$$H_N = H_1^p + \sum_{j=p+1}^{N} h(i) + \sum_{i=1}^{p} \sum_{j=p+1}^{N} g(i,j) + \sum_{i<j(i\geq p+1)}^{N} g(i,j), \tag{7}$$

where

$$H_1^p = \sum_{i=1}^{p} h(i) + \sum_{i<j}^{p} g(i,j), \tag{8}$$

and apply the contraction operator $L_N^p$ on both sides of the Eq. (3), we develop an equation that the $p$th-order density operator satisfies [4]

$$\begin{aligned} -\frac{\partial}{\partial \beta} D^p \;=\; & H_1^p D^p + (N-p) L_{p+1}^p \left[ h(p+1) D^{p+1} \right] + (N-p) L_{p+1}^p \left[ \sum_{i=1}^{p} g(i, p+1) D^{p+1} \right] + \\ & + \binom{N-p}{2} L_{p+2}^p \left[ g(p+1, p+2) D^{p+2} \right]. \end{aligned} \tag{9}$$

It provides a law according to which the reduced density operators vary in terms of the change of temperature.

### 3.2 Hierarchy Bloch Equations for Reduced Density Operators in Grand Canonical Ensemble

The above scheme for deducing the equations for the reduced operators can be readily extended to the case of a grand canonical ensemble, which is a more general one with a fluctuating particle number $N$. In this ensemble, the density operator is defined in the entire *Fock* space

$$F = \sum_{N=0}^{\infty} \oplus V^N,$$

and is written as the direct sum of the density operators $D_G(N)$ associated with the $N$-particle *Hilbert* space $V^N$,

$$D_G = \sum_{N=0}^{\infty} \oplus D_G(N), \tag{10}$$

where

$$\begin{aligned} D_G(N) &= \exp[-\beta(H - \mu N)], \\ &= \exp(-\beta \bar{H}), \end{aligned} \tag{11}$$

and

$$\bar{H} = H - \mu N, \tag{12}$$

is called the grand *Hamiltonian* on $V^N$. The form of the *Hamiltonian* $H$ has been given by Eq. (4) and the $\mu$ is the chemical potential. The corresponding $p$th-order reduced density operator is therefore defined as

$$D_G^p = \sum_{N=p}^{\infty} \oplus \begin{pmatrix} N \\ p \end{pmatrix} L_N^p[D_G(N)], \tag{13}$$

with the trace given by

$$Tr(D_G^p) = \left\langle \begin{pmatrix} N \\ p \end{pmatrix} \right\rangle D_G^0 \tag{14}$$

and

$$D_G^0 = \Xi(\beta, \mu, V). \tag{15}$$

The $\Xi(\beta, \mu, V)$ is the grand partition function.

In a similar manner, we can also derive the hierarchy equations that the reduced density operators in the grand canonical ensemble obey [5]

$$\begin{aligned} -\frac{\partial}{\partial \beta} D^p &= \bar{H}_1^p D^p + (p+1) L_{p+1}^p \left[ \bar{h}(p+1) D^{p+1} \right] + (p+1) L_{p+1}^p \left[ \sum_{i=1}^{p} g(i, p+1) D^{p+1} \right] + \\ &+ \begin{pmatrix} p+2 \\ 2 \end{pmatrix} L_{p+2}^p \left[ g(p+1, p+2) D^{p+2} \right], \end{aligned} \tag{16}$$

where

$$\bar{H}_1^p = \sum_{i=1}^{p} \bar{h}(i) + \sum_{i<j}^{p} g(i,j), \tag{17}$$

and

$$\bar{h}(i) = h(i) - \mu. \tag{18}$$

It gives us a law with which the reduced density operators in the grand canonical ensemble vary in temperature.

### 3.3 Orbital Approximation and Hartree-Fock Equation at Finite Temperature

The Eqs. (9) and (16) define a set of *hierarchy* equations that establish the relation among the reduced density operators $D^p$, $D^{p+1}$, and $D^{p+2}$. They can be solved either in an exact scheme or by an approximate method. The previous study of $N$ electrons with an independent particle approximation to the *Schrödinger* equation for their *pure* states has lead to the *Hartree – Fock* equation for the molecular orbitals [94-100]. We thereby expect that the same approximate scheme to the reduced *Bloch* equations (9) or (16), which hold for more general mixed states, will yield more generic eigenequations than the usual *Hartree – Fock* equation for the molecular orbitals.

We consider the case of a grand canonical ensemble. When $p = 1$, Eq. (16) reads

$$-\frac{\partial}{\partial\beta}D^1 = \bar{H}_1 D^1 + \frac{Tr(\bar{h}D^1)}{D^0}D^1 - \frac{1}{D^0}D^1\bar{h}D^1 + 2L_2^1\left[g(1,2)D^2\right] + 3L_3^1\left[g(2,3)D^3\right]. \tag{19}$$

Under the orbital approximation, the above second-order and third-order reduced density operators for the electrons can be written as

$$D^3 = D^1 \wedge D^1 \wedge D^1/(D^0)^2 \tag{20}$$

and

$$D^2 = D^1 \wedge D^1/D^0. \tag{21}$$

These are the special situations for the statement that a $p$th-order reduced density matrix can be expressed as a $p$-fold *Grassmann* product of its first-order reduced density matrices. With these approximations, the last two terms of Eq. (19) can be evaluated in a straightforward way as follows

$$2L_2^1\left[g(1,2)D^2\right] = (J-K)D^1, \tag{22}$$

and

$$3L_3^1\left[g(2,3)D^3\right] = \frac{Tr(gD^2)}{D^0} - \frac{1}{D^0}D^1(J-K)D^1, \tag{23}$$

where

$$J = Tr_2\left[g \cdot D^1(2;2)\right]/D^0, \tag{24}$$

and

$$K = Tr_2 \left[ g \cdot (2,3) \cdot D^1(2;2) \right] / D^0, \tag{25}$$

are called the *Coulomb* and exchange operators, respectively. With the $(2,3)$ being the exchange between the particle 2 and 3, the action of the $K$ on the reduced density operator is

$$\begin{aligned} K \cdot D^1(3;3) &= Tr_2 \left[ g \cdot (2,3) \cdot D^1(2;2) \right] / D^0 \cdot D^1(3;3) \\ &= Tr_2 \left[ g \cdot D^1(3;2) \cdot D^1(2;3) \right] / D^0. \end{aligned} \tag{26}$$

Substitution of Eqs. (22) and (23) into Eq. (19) yields the *Bloch* equation for the first-order reduced density matrix of $N$ interacting electrons under orbital approximation,

$$-\frac{\partial}{\partial \beta} D^1 = (F - \mu) D^1 + \left( \frac{Tr\bar{h} D^1}{D^0} + \frac{Tr D^2}{D^0} \right) D^1 - \frac{1}{D^0} D^1 (F - \mu) D^1, \tag{27}$$

where

$$F = h + J - K, \tag{28}$$

is called the *Fock* operator at finite temperature. Redefine the normalized first-order reduced density operator

$$\rho^1 = D^1 / D^0, \tag{29}$$

we can simplify above equation into

$$-\frac{\partial}{\partial \beta} \rho^1 = (F - \mu) \rho^1 - \rho^1 (F - \mu) \rho^1. \tag{30}$$

Furthermore, from Eq. (30) and its conjugate, we get

$$F \rho^1 - \rho^1 F = 0, \tag{31}$$

which means that the *Fock* operator $F$ and the first-order reduced density matrix $\rho^1$ commute. They are also *Hermitian*, and therefore have common eigenvectors $\{|\phi_i\rangle\}$. These vectors are determined by the following eigen equation for the *Fock* operator,

$$F |\phi_i\rangle = \epsilon_i |\phi_i\rangle. \tag{32}$$

The first-order reduced density operator is correspondingly expressed as

$$\rho^1 = \sum_i \omega(\beta, \mu, \epsilon_i) |\phi_i\rangle \langle \phi_i|, \tag{33}$$

where $\omega(\beta, \mu, \epsilon_i)$ is the thermal probability that the orbital is found to be in the state $\{|\phi_i\rangle\}$ at finite temperature $T$. Substituting Eq. (33) into Eq. (30), we can obtain the equation this thermal probability $\omega(\beta, \mu, \epsilon_i)$ satisfies,

$$-\frac{\partial}{\partial \beta} \omega(\beta, \mu, \epsilon_i) = (\epsilon_i - \mu) \omega(\beta, \mu, \epsilon_i) - (\epsilon_i - \mu) \omega^2(\beta, \mu, \epsilon_i). \tag{34}$$

Its solution has the same usual form of the *Fermi* − *Dirac* statistics for the free electron gas as follows,

$$\omega(\beta, \mu, \epsilon_i) = \frac{1}{1 + e^{\beta(\epsilon_i - \mu)}}, \tag{35}$$

with the energy levels $\{\epsilon_i\}$ determined by Eq. (32).

## 4 Discussions and Prospects

In this paper, we have presented a description of both experimental and theoretical developments related to the temperature impact on the microscopic structure and processes for the molecules.

In the theoretical part of this paper, we have depicted the sets of hierarchy *Bloch* equations for the reduced statistical density operators in both a canonical and a grand canonical ensembles for the identical fermion system with two-body interaction. We have solved the equations in the latter case under a single-orbital approximation and obtained an eigen-equation for the single-particle states. It is the extension of usual commonly used *Hartree* − *Fock* equation at the absolute zero temperature to the situation at any finite temperature. The average occupation number formula for each single-particle state is also obtained, which has the same analytical form as that for the free electron gas with the single-particle state energy determined by the *Hartree* − *Fock* equation at finite temperature (32).

¿From Eqs. (24), (25) and (28), we see that the *Coulomb* operator $J$, the exchange operator $K$, and therefore the *Fock* operator $F$ are both the coherent and the incoherent superpositions of single-particle states. They are all temperature-dependent through an incoherent superposition factor, the *Fermi* − *Dirac* distribution, $\omega(\beta, \mu, \epsilon_i)$. Therefore, the mean force or the force field, and the corresponding microscopic structure are temperature-dependent.

We have expounded the physics relating to the temperature effects on the electronic structure or processes of the molecules. This is very critical for our understanding and studying the temperature influence over the molecular structure. From this analysis, for example, we could see that the temperature should have a stronger effect on the molecular transition states and therefore their chemical reactivity. Starting from Eqs. (9) or (16), it will be a very significant work to establish a corresponding multireference theory for the molecular orbitals at finite temperature [96, 101, 102].

On the experimental sides, we have exposed four major fields of investigations of chemical or biomolecular interests, which show the temperature impact on their structures, spectra, or bond rupture forces.

The complete determination of the geometric structure and electronic spectra of the polymeric molecules is a very difficult task. As has been stated in papers [15, 16], there are many different components contributing to the change of the spectra. At present, we focus on the study of the effect from the temperature. We have demonstrated that it can alter both the shapes and positions of the absorption and other spectra for the *PPV* and its derivatives. As has been analyzed, the increase of the temperature will cause the excitation of the vibrational, rotational and liberal motions which might also lead to the electronic transition. The $Huang - Rhys$ parameter has been introduced to describe the strength of the coupling between the electronic ground- and excited-state geometries. Furthermore, it has been found that this factor is an increasing function of the temperature [7, 8, 9]. Obviously, a more detailed study of the electronic structure, excitation and spectroscopic signature at the first-principle level which includes the temperature-dependent force field is expected.

The temperature has proved to be a big player in both experimental and theoretical study of the structure and dynamics of biomolecules including their folding or unfolding. At a first glance, the energy gap between the $HOMO$ and $LUMO$ for the biomolecules should be small or comparable to the $Boltzmann$ thermal energy $k_BT$ because of their very large molecular size. Therefore a change of the temperature should have a strong influence over their electronic states, and consequently, the energy landscape and the related dynamics including the folding or unfolding, etc. The experimental investigation with the $AFM$ and other techniques of the temperature effect on the shift of their unfolding pathways might have verified this sort of thermal deformation of potential energy landscape [19, 20, 21]. This is in contrast to the tilt and deformation of energy landscape of biomolecules including their transition states resulting from the applied mechanical forces [69].

Unfolding proteins by temperature is not just one of the classical experimental techniques for the study of the structure, dynamics and energetics of the biomolecules. It has also been employed, for example, in the molecular dynamic simulation to study the structure of transition states of $CI2$ in water at two different temperatures: 298 $K$ and 498 $K$ [22]. The later high temperature is required in order to destabilize the native state for monitoring the unfolding as done in the real experiments. In another recent molecular dynamics simulation [23], $Karplus$'s group has compared the temperature-induced unfolding with the force-stretching unfolding for two $\beta$-sandwich proteins and two $\alpha$-helical proteins. They have found that there are the significant differences in the unfolding pathways for two approaches. Nevertheless, in order to get more reliable results, temperature-dependent force fields need to be developed and included in the molecular dynamics simulations. This is also the case in the theoretical investigation of protein folding since an accurate simulation of protein folding pathways requires better stochastic or temperature-dependent potentials which have become the bottleneck in structure prediction [24, 25, 26, 27, 28]. All in all, the variation of temperature leads to the change of the mean force or the energy landscape, and therefore provides a vast variety of possibilities, for instance, in the protein design and engineering.

The intermolecular forces are ubiquitous in nature. They are also extremely important for the biological systems and for the existence of life. The intermolecular forces have the specificity which are responsible for the molecular recognition between receptor and ligand, antibody and antigen, and complementary strands of $DNA$, and therefore for the regulation of complex organization of life [103]. For these reasons, the experiment carried out in $Beebe$'s group has an immediate significance. It has demonstrated that the temperature can be an important factor for changing the specificity of the intermolecular forces and therefore the function of life [63]. Nevertheless, how the charge redistribution occurs due to the variation of the temperature has not been interpreted, and a microscopic theory for quantifying the temperature influence on the intermolecular forces is still lacking. Since the delicate study of the intermolecular forces provides the insight into complex mechanisms of ligand-receptor binding and unbinding processes or pathways, a paramount future research is to establish the links between the intermolecular forces and the temperature within the quantum many-body theory.

The theoretical study of the temperature effects on the optical absorption spectra of solvated electron is still in very early stage and few published works are available [82, 83, 84, 85]. One of the earliest studies by $Jortner$ used a cavity model to simulate the solvated electron where the electron is confined to the cavity surrounded by the dielectric continuum solvent [82]. However, his study is not of fully microscopic in nature since he assumed a temperature dependence of phenomenological dielectric constants which were also obtained from the available experimental data. In addition, the model used is too simplified and, for instance, it neglects the intrinsic structure of the solvent molecules. There are a few recent investigations on the temperature effects on the absorption spectra of solvated electrons. They all cannot catch the full features of the experimental observations. One reason is that the physical nature for the process is not totally understood which might leads to incorrect models used for the simulation. The other is to utilize the crude models which might have omitted some important physical effects. For example, in an analysis by $Brodsky$ and $Tsarevsky$ [83], they have concluded a temperature-dependence relation for the spectra which is, however, in contradiction with the experimental findings at high temperature. The quantum path-integral molecular dynamics simulation cannot produce those temperature-dependence relations observed in the experiments [84]. In a recent quantum-classical molecular-dynamics study by $Nicolas\ et\ al$, even though the temperature-dependent features of optical absorption spectra for the solvated electron in water have been recovered [85], however, they claim that the red shifts of absorption spectra with increasing temperature observed in both experiments and calculation are due to the density effect instead of temperature. This might cast the doubt of usefulness of our current theoretical work in this area. However, after examining their work, we observe that they actually have $not$ included any temperature effect on the electron in their theoretical model. This effect might be either from the $Fermi - Dirac$ distribution for individual electrons or due to the electronic excitation caused by the thermal excitation of the solvent, as we have discussed in paper [4]. Obviously, much finer theoretical work or more experimental investigations in this area are expected to resolve this dispute.

In addition to the systems discussed above, there are other types which also show the temperature impact on their microscopic structures. Either theoretical or experimental work have been done or are in progress. Examples include the study of the temperature dependence of the $Coulumb$ gap and the density of states for the $Coulomb$ glass, the experimental investigation of the temperature effects on the band-edge transition of $ZnCdBeSe$, and the theoretical description of the influence from the temperature on the polaron band narrowing in the oligo-acene crystals [104, 105, 106].

To sum up, the quantum chemistry at finite temperature is a new and exciting field. With the combination of the techniques from current quantum chemistry with those developed in statistical or solid state physics, it will provide us a myriad number of opportunities for the exploration.

## References


[1] Heitler W.; London F. Z. *Physik* 1927, 44, 455-472.

[2] Wei L. (2003). Exploring the Harmony between Theory and Computation - toward a unified electronic structure theory. arXiv: physics/0307156.

[3] Wei L.; Sun C. C.; Zeng Z. H. *Ann. Physics* (2004), online available.

[4] Wei L.; Sun C. C. *Physica A* 2004, 334, 144-150.

[5] Wei L.; Sun C. C. *Physica A* 2004, 334, 151-160.

[6] Löwdin P.-O. *Int. J. Quantum Chem.* 1986, 29, 1651-.

[7] Hagler T. W.; Pakbaz K; Voss K. F.; Heeger A. *J. Phys. Rev.* B 1991, 44, 8652-8666.

[8] Pichler K.; Halliday D. A.; Bradley D. D. C.; Burn P. L.; Friend R. H.; Holmes A. B. *J. Phys.: Condens. Matter* 1993, 5, 7155-7172.

[9] Yu J.; Hayashi M.; Lin S. H.; Liang K.-K.; Hsu J. H.; Fann W. S.; Chao C.-I.; Chuang K.-R.; Chen S.-A. *Synth. Met.* 1996, 82, 159-166.

[10] Spiegelberg Ch.; Peyghambarian N.; Kippelen B. *App. Phys. Lett.* 1999, 75, 748-750.

[11] Sheridan A. K.; Lupton J. M.; Samuel I. D. W.; Bradley D. D. C. *Chem. Phys. Lett.* 2000, 322, 51-56.

[12] Peeters E.; Ramos A. M.; Meskers S. C. J.; Janssen R. A. J. *J.Chem. Phys*. 2000, 112, 9445-9454.

[13] Gierschner J., Mack H.-G.; Lüer L.; Oelkru D; *J. Chem. Phys.* 2002, 116, 8596-8609.

[14] Yu J.; Fann W. S.; Lin S. H. *Theor. Chem. Acc.* 2000, 103, 374-379.

[15] Kwasniewski S. P.; Francois J. P.; Deleuze M. S.; *Inter. J. Quantum Chem.* 2001, 85, 557-568.

[16] Kwasniewski S. P.; Francois J. P.; Deleuze M. S., *J. Phys. Chem. A* 2003, 107, 5168-5180.

[17] Rakow A. L.; Hochmuch R. *Biophys. J.* 1975, 15, 1095-1100.

[18] Waugh R.; Evans E. A. *Biophys. J.* 1979, 26, 115-132.

[19] Clausen-Schaumann H.; Rief M.; Tolksdorf C.; Gaub H. E. *Biophys. J.* 2001, 78, 1997-2007.

[20] Williams M. C.; Wenner J. R.; Rouzina I.; Bloomfield V. A. *Biophys. J.* 2001, 80, 1932-1939.

[21] Law R.; Liao G.; Harper S.; Yang G.; Speicher D. W.; Discher D. E. *Biophys. J.* 2003, 85, 3286-3293.

[22] Li A.; Daggett V. J. *Mol. Biol.* 1996, 257, 412-429.

[23] Paci E.; Karplus M.; *Proc. Natl. Acad. Sci.* 2000, 97, 6521-6526.

[24] Brynegelson J. D.; Wolynes P. G. J. *Phys. Chem.* 1989, 93, 6902-6915.

[25] Shakhnovich E. I.; Gutin A. M. *Biophys. Chem.* 1989, 34, 187-199.

[26] Karplus M.; Šali A.; *Curr. Opin. Struct. Biol.* 1995, 5, 58-73.

[27] Dill K. A. et al. *Protein Sci.* 1995, 4, 561-602.

[28] Thirumulai D.; Woodson S. A. *Acc. Chem. Res.* 1996, 29, 433-439.

[29] Minetti M.; Ceccarini M.; Stasi A. Maria M. Di; Petrucci T. C.; Marchesi V. T. J. *Cell. Biochem.* 1986, 30, 361-370.

[30] Grimaldi P.; Minetti M.; Pugliese P.; Isacchi G. J. *Cell. Biochem.* 1989, 41, 25-35.

[31] Binnig G.; Quate C. F.; Gerber Ch. *Phys. Rev. Lett.* 1986, 56, 930-933.

[32] Binnig G.; Rohrer H. *Rev. Mod. Phys.* 1987, 59, 615-625.

[33] Hansma P. K.; Ellings V. B.; Marti O.; Bracker C. E. *Science* 1988, 242, 209-216.

[34] Drake B.; Prater C. B.; Weisenhorn A. L.; Gould S. A. C.; Albrecht T. R.; Quate C. F.; Cannell D. S.; Hansma H. G.; Hansma P. K. *Science* 1989, 243, 1586-1589.

[35] Radmacher M.; Tillmann R. W.; Fritz M.; Gaub H. E. *Science* 1992, 257, 1900-1905.

[36] Moy V. T.; Florin E.-L.; Gaub H. E. *Colloids Surfaces* 1994, 93, 343-348.

[37] Florin E.-L.; Moy V. T.; Gaub H. E. *Science* 1994, 264, 415-417.

[38] Moy V. T.; Florin E.-L.;, Gaub H. E. *Science* 1994, 266, 257-259.

[39] Lo Y.-S.; Huefner N. D.; Chan W. S.; Stevens F.; Harris J. M.; Beebe T. P. *Jr. Langmuir* 1999, 15, 1373-1382.

[40] Han T.; Williams J. M.; Beebe T. P. *Jr. Anal. Chim. Acta.* 1995, 307, 365-376.

[41] MacDonald R. I.; Pozharski E. V. *Biochemistry* 2001, 40, 3974-3984.

[42] Rief M.; Gautel M.; Oesterhelt F.; Fernandez J. M.; Gaub H. E. *Science* 1997, 276, 1109-1112.

[43] Carrion-Vazquez M.; Oberhauser A. F.; Fowler S. B.; Marszalek P. E.; Broedel S. E.; Clarke J.; Fernandez J. M. *Proc. Natl. Acad. Sci.* USA 1999, 96, 3694-3699.

[44] Lu H.; Isralewitz B.; Krammer A.; Vogel V.; Schulten K.; *Biophys. J.* 1998, 75, 662-671.

[45] Carl P.; Kwok C. H.; Manderson G.; Speicher D. W.; Discher D. E. *Proc. Natl. Acad. Sci.* USA 2001, 98, 1565-1570.

[46] Oberhauser A. F.; Marszalek P. E.; Erickson H. P.; Fernandez J. M. *Nature* 1998, 393, 181-185.

[47] Lenne P.-F.; Raae A. J.; Altmann S. M.; Saraste M.; Horber J. K. H. *FEBS Lett.* 2000, 476, 124-128.

[48] Law R.; Carl P.; Harper S.; Dalhaimer P.; Speicher D. W.; Discher D. E. *Biophys. J.* 2003, 84, 533-544.

[49] Rouzina I.; Bloomingfield V. A. *Biophys. J.* 2001, 80, 882-893.

[50] Rouzina I.; Bloomingfield V. A. *Biophys. J.* 2001, 80, 894-900.

[51] Grandbios M.; Beyer M.; Rief M.; Clausen-Schaumann H.; Gaub H. E. *Science* 1999, 283, 1727-1730.

[52] Jarvet J.; Damberg P.; Danielsson J.; Johansson I.; Eriksson L. E. G.; Gräslund A. *FEBS Letters* 2003, 555, 371-374.

[53] Xue H.; Guo R.; Wen Y.; Liu D.; Huang L. *J. Bacteriol.* 2000, 182, 3929-3933.

[54] Tong Y.; Cui Q.; Feng Y.; Huang L.; Wang J. *J. Biomol. NMR* 2002, 22, 385-386.

[55] Cui Q.; Tong Y.; Xue H.; Huang L.; Feng Y.; Wang J. *J. Biol. Chem* 2003, 278, 51015-51022.

[56] Rist W.; Jorgensen T. J. D.; Roepstorff P.; Bukau B.; Mayer M. P. *J. Biol. Chem.* 2003, 278, 51415-51421.

[57] Vu D. M.; Myers J. K.; Oas T. G.; Dyer R. B. *Biochemistry* 2004, 43, 3582-3589.

[58] Green N. M. *Adv. Prot. Chem.* 1975, 29, 85-134.

[59] Darst S. A. et al. *Biophys. J.* 1991, 59, 387-396.

[60] Livnah O.; Bayer E. A.; Wilchek M.; Sussman J. L. *Proc. Natl. Acad. Sci.* USA 1993, 90, 5076-5080.

[61] Pugliese L.; Coda A.; Malcovati M.; Bolognesi M. *J. Mol. Biol.* 1993, 231, 698-710.

[62] Pugliese L.; Coda A.; Malcovati M.; Bolognesi M. *J. Mol. Biol.* 1994, 235, 42-46.

[63] Lo Y.-S.; Simons J.; Beebe T. P. *Jr.J. Phys. Chem.* B 2002, 106, 9847-9852.

[64] Williams J. M.; Han T.; Beebe T. P. *Jr. Langmuir* 1996, 12, 1291-1295.

[65] Wenzler L. A.; Moyes G. L.; Olson L. G.; Harris J. M.; Beebe T. P. *Jr. Anal. Chem.* 1997, 69, 2855-2861.

[66] Wei Z. Q.; Wang C.; Zhu C. F.; Zhou C. Q.; Xu B.; Bai C. L. *Surf. Sci.* 2000, 459, 401-412.

[67] Grubmüller H.; Heymann B.; Tavan P. *Science* 1996, 271, 997-999.

[68] Izrailev S.; Stepaniants S.; Balsera M.; Oono Y.; Schulten K. *Biophys. J.* 1997, 72, 1568-1581.

[69] Evans E.; Ritchie K. *Biophys. J.* 1997, 72, 1541-1555.

[70] Evans E. *Annu. Rev. Biophys. Biomol. Struct.* 2001, 30, 105-128.

[71] Blades H.; Hodgins J. W. Can. *J. Chem.* 1955, 33, 411-425.

[72] Gottschall W. C.; Hart E. J. *J. Phys. Chem.* 1967, 71, 2102-2106.

[73] Michael B. D.; Hart E. J.; Schmidt K. H. *J. Phys. Chem.* 1971, 75, 2798-2805.

[74] Jou F.-Y.; Freeman G. R. *J. Phys. Chem.* 1979, 83, 2383-2387.

[75] Christensen H.; Sehested K. *J. Phys. Chem.* 1986, 90, 186-190.

[76] Shiraishi H.; Katsumura Y.; Hiroishi D.; Ishigure K.; Washio M. *J. Phys. Chem.* 1988, 92, 3011-3017.

[77] Herrmann V.; Krebs P. *J. Phys. Chem.* 1995, 99, 6794-6800.

[78] Chandrasekhar N.; Krebs P., *J. Chem. Phys.* 2000, 112, 5910-5914.

[79] Wu G.; Katsumura Y.; Muroya Y.; Li X.; Terada Y.; *Chem. Phys. Lett.* 2000, 325, 531-536.

[80] Mostafavi M.; Lin M.; Wu G.; Katsumura Y.; Muroya Y.; *J. Phys. Chem. A* 2002, 106, 3123-3127.

[81] Mostafavi M.; Lin M.; He H.; Muroya Y.; Katsumura Y. *Chem. Phys. Lett.* 2004, 384, 52-55.

[82] Jortner J. *J. Chem. Phys.* 1959, 30, 839-846.

[83] Brodsky A.M.; Tsarevsky A.V. *J. Phys. Chem.* 1984, 88, 3790-3799.

[84] Wallqvist A.; Martyna G.; Berne B. J. *J. Phys. Chem.* 1988, 92, 1721-1730.

[85] Nicolas C.; Boutin A.; Lėvy B.; Borgis D. *J. Chem. Phys.* 2000, 118, 9689-9696.

[86] Drickamer H. G.; Franck C. W. Electronic Transitions and the High Pressure Chemistry and Physics of Solids; London, Chapman and Hall: New York, NY, 1973.

[87] Drickamer H. G. *Annu. Rev. Phys. Chem.* 1982, 33, 25-47.

[88] Hartree D. R. *Proc. Camb. Phil. Soc.* 1928, 24, 111-132.

[89] Fock V. Zeit. *Physick* 1930, 61, 126-148.

[90] Bloch F. Zeits. f. *Physick* 1932, 74, 295-335.

[91] Kirkwood J. G. *Phys. Rev.* 1933, 44, 31-37.

[92] Kummer H. *J. Math. Phys.* 1967, 8, 2063-2081.

[93] Harriman J. E. *Phys. Rev. A* 1978, 17, 1257-1268.

[94] Cohen L.; Frishberg C. *Phys. Rev. A* 1976, 13, 927-930.

[95] Nakatsuji H. *Phys. Rev. A* 1976, 14, 41-50.

[96] Schlosser H. *Phys. Rev. A* 1977, 15, 1349-1351.

[97] Harriman J. E. *Phys. Rev. A* 1979, 19, 1893-1895.

[98] Valdemoro C. *Phys. Rev. A* 1992, 45, 4462-4467.

[99] Mazziotti D. A. *Phys. Rev. A* 1998, 57, 4219-4234.

[100] Coleman A. J. *Intern. J. Quantum Chem.* 2001, 85, 196-203.

[101] Wahl A. C.; Das G. *J. Chem. Phys.* 1972, 56, 1769-1775.

[102] Gilbert T. L. *J. Chem. Phys.* 1974, 60, 3835-3844.

[103] Frauenfelder H.; Parak F.; Young R. D. *Annu. Rev. Biophys. Biophys. Chem.* **17**, 451-479 (1988).

[104] M. Sarvestani, M. Schreiber, and T. Vojta, *Phys. Rev. B* **52**, R3820-3823 (1995).

[105] Hsien C. H.; Huang Y. S.; Ho C. H.; Tiong K. K.; Munoz M.; Maksimov O.; Tamargo M. C. Japn. *J. App. Phys. Part 1* 2004, 43, 459-466.

[106] Hannewald K.; Stojanovic V. M.; Bobbert P. A. *J. Phys.: Condens. Matt.* 2004, 16, 2023-2032.

In: Progress in Chemical Physics Research
Editor: A. N. Linke, pp. 21-64

ISBN 1-59454-451-4


*Chapter 2*

# TEMPERATURE-INDUCED TRANSITIONS IN FREE CLUSTERS

***F. Calvo****
Laboratoire de Physique Quantique,
IRSAMC, Université Paul Sabatier,
118 Route de Narbonne, F31062 Toulouse Cedex

**Abstract**

The phenomenology of phase transitions in free atomic and molecular clusters is reviewed, based on how it is revealed from numerical simulations and experiments. We mainly consider transitions involving condensed phases, paying a particular attention at the finite size analog of melting as well as solid-solid transitions. The chosen examples range from weakly bound van der Waals clusters to ionic or metal clusters. In general we attempt to correlate the observed behavior to the features of the energy landscape, and possible indicators to characterize phase changes are critically discussed. The transition to the vapor phase is mentioned with a view at evaporation phenomena, and a way to quantify the cluster boiling point is proposed.

**Keywords:** *atomic clusters, structure, melting, boiling, stability.*

## 1 Introduction

Cluster thermodynamics has been the subject of an intense work in the past two decades, mostly theoretically but also, more recently, experimentally. This Chapter intends to provide a subjective overview of some salient results obtained from numerical or real experiments on this topic. Clusters are finite size aggregates containing between a few and tens of thousands of atoms or molecules. Their geometric, electronic, and magnetic properties are

*E-mail address: florent@irsamc.ups-tlse.fr

known to depend significantly on the chemical species involved, the physical environment, as well as size itself [1, 2]. In some cases, the nature of chemical bonding is also a function of system size. Mercury clusters, for instance, exhibit van der Waals interactions up to the size of about 13, then covalent bonding, before eventually becoming metallic after several tens of atoms.

The peculiar properties of clusters are frequently explained by their large surface/volume ratio, which makes them very appealing, e.g., in catalysis. Large clusters are often considered as small pieces of bulk matter. This point of view is associated with condensed matter approaches and simplified Hamiltonians, and coarse-grained asymptotic descriptions such as the liquid drop model from nuclear physics. Here a given property $P(n)$ is empirically expanded into a short series containing the volume and surface contributions, and possibly extra terms, expressed as a power law in $n^{-1/3}$. The associated behavior of $P(n)$ is usually monotonic. On the other hand, accurate quantum chemical calculations yield highly detailed information about the behavior of small clusters at the scale of a single atom or molecule. These "cluster size effects" [3] are then likely to be characterized by non monotonic variations with increasing size.

Clusters provide the link between the atom and the bulk. Beyond this common belief, thermodynamical properties naturally pertain to the macroscopic world, because one would easily think of them not to have any counterpart at the molecular level, unlike cohesion energies or magnetic moments, for instance. Statistical mechanics are a rigorous and general framework to treat finite systems in the microcanonical and canonical ensembles. However the actual situation of free aggregates presents some difficulties related to their finite size character. First, if a cluster has enough internal energy to dissociate, then it will, under time scales easily reached in experiments, typically the microsecond or less. This is an important matter for small systems, whose properties are very sensitive to their size [4]. In addition, the notions of phases and phase transitions are not obvious when dealing with clusters. Even the well known definition of a liquid, namely something which takes the shape of its container, can obviously not be extended to clusters. Finally, the volume of the system and its conjugate thermodynamical variable, the pressure, also become loose concepts, but they are crucial for a correct description of the possible vapor phase of finite systems and the related phase transitions.

Historically, the interest in cluster thermodynamics arose in liquid-gas nucleation theory. In the early Monte Carlo (MC) simulations by Abraham and coworkers [5], the focus was not on the clusters themselves, but instead on their equilibrium with the external gas. During the 1970s, some molecular dynamics (MD) simulations by McGinty [6] and by Briant and Burton [7] among others, uncovered the first hints of a phase change in a small cluster by calculating the caloric curve, that is the average temperature versus total energy. However, MC calculations carried out shortly after this work by Kaelberer and Etters [8], Nauchitel and Persin [9], and by Quirke and Sheng [10] were not in full agreement with the MD results. These discrepancies were then attributed to the possible lack of statistics and to the different methods used.

With the great improvement in computer power available in the following decade, Berry and coworkers brought decisive contributions to the field, especially with the discovery of the dynamical coexistence phenomenon [11]. When in the phase of dynamical coexistence, a cluster remains in its solid phase for some time, and then switches to its liquid phase, before going back to the solid phase, and so on. In the simple Lennard-Jones (LJ) clusters studied by these authors, each phase was seen to be homogeneous (fully solid, or fully liquid), which was later interpreted as the consequence of the too high energetic cost of creating a solid/fluid interface in a small system. Berry *et al.* also generalized the use of specific observables or indicators to characterize the mechanical or thermodynamical state of the cluster [12, 13, 14, 15, 16]. For example, the root mean square bond length fluctuation or Lindemann index, remains one of the most convenient quantities for characterizing the solid and liquid states in numerical simulations. The work by Berry and coworkers helped in interpreting the disagreement between early simulations, by noting that MC and MD results should *not* be the same since the statistical ensembles to which they pertain, canonical and microcanonical, respectively, are not strictly identical far below the bulk limit.

Further statistical grounds for understanding cluster thermodynamics were added by Labastie and Whetten [17], who elucidated the nature of melting in these finite systems. The change between solid and liquid was interpreted from the caloric curves to be a first order transition rounded by size effects: the sharp, delta curve of the bulk heat capacity is replaced by a broad peak. Simultaneously, and pursuing previous efforts by the Berry group, Wales introduced another way of investigating cluster thermodynamics, by looking at their energy landscape — as it would soon be called. The importance of structure was recognized before in the relationship between dynamical coexistence and the microcanonical caloric curve [14], where the occurence of a S-bend in the kinetic temperature was shown to result from a dynamical equilibrium between a hot solid phase and a colder liquid phase.

In the 1990s, a great variety of atomic and molecular clusters became the subject of simulation studies concerned with temperature effects. While the LJ potential still provides a very useful model nowadays, especially for developing and refining theoretical tools, weakly bound molecular clusters, ionic clusters, covalent or metal clusters could be investigated in wide ranges of thermodynamical conditions. The advent of many new numerical or semi-analytical methods formed a fairly complete framework to predict and understand the thermal behavior of clusters, in their condensed phases. In addition, significant advances in global optimization techniques and better recipes to deal with broken ergodicity allowed larger clusters and longer time scales to be reached by computers.

The numerous, sometimes surprising properties of clusters induced by a finite temperature lead several experimental groups to address phase transition problems. The influence of temperature on the spectroscopic properties of van der Waals clusters doped with a chromophore molecule [18, 19] did not provide full evidence of melting in such small clusters. Later, Haberland and coworkers failed to measure the melting point in size-selected sodium clusters from their absorption spectrum, the observed effect being too small [20].

Indirect measurements of the melting point of a free cluster was obtained by Martin and coworkers [21] by looking at the electron diffraction pattern, and the disappearance of the Bragg peaks above some characteristic temperature. This method is basically the same as used previously by Takagi [22] on thin films, and by Buffat and Borel on deposited nanoparticles [23]. Contrary to the spectroscopic approach, the electron diffraction technique requires clusters that are large enough to yield cristallographic features. Ion chromatography measurements by the Jarrold group were used to get structural information on charged clusters, and more recently to study rearrangements kinetics [24] and even melting points [25].

A breakthrough was accomplished in 1997 by the group of Haberland [26], who inferred the complete caloric curve of the cluster $Na_{139}^+$ from photofragmentation spectra. These authors deduced the melting point and latent heat of fusion from the shape of the caloric curves, and repeated their measurements in a wide range of cluster sizes. The thermal properties were seen to exhibit strong, non monotonic size effects [27]. Building upon their initial experimental setup, this group also provided experimental support for the dynamical coexistence phenomenon in $Na_{147}^+$ [28], but also for the existence of a boiling point evidenced by a plateau in the caloric curve [29].

The search for critical phenomena associated with the liquid-gas transition was motivated by similar work in nuclear physics [30]. The caloric curve of highly energetic clusters has been reported to exhibit some features interpreted as the signature of the liquid-gas transition. This has been shown for charged strontium [31] and hydrogen [32] clusters, even though the interpretation of the latter measurements has been debated [33]. Since the work by Klots [34], the stability of hot clusters is known to be highly sensitive to the observation time. To deal with clusters as they would be found in free jet expansions, Klots introduced the "evaporative ensemble" in which the average size of the cluster results from a dynamical equilibrium between previous and future evaporations. The evaporative ensemble is somewhat related to the grand canonical ensemble of statistical mechanics, because the system size is not strictly fixed. However it explicitely includes the time variable.

Temperature effects in clusters are one of the main focus of their thermodynamical study, but there are other situations where they should not be neglected. At moderate temperatures, when the cluster remains in a single isomer acting as a solid phase, spectroscopic measurements are perturbed by a broadening and a shift of their signal induced by temperature. As we shall see below, quantum effects may also be important, and should sometimes be accounted for in the theoretical interpretation. Actually, any property related to cluster structure (which basically means most of them) are likely to undergo some changes as the cluster is heated.

Increasing its temperature, the cluster may have access to new isomers higher in energy. There may be only very few of them, but separated in configuration space by large energy barriers. Upon suitable conditions, solid-solid transitions may arise. This situation will be illustrated in section 2. The melting point of the cluster is characterized by the appearance of a large bunch of new isomers. Accross melting and in the liquidlike phase, temperature can obviously play a very important role, especially for spectroscopic properties, which are highly dependent on molecular conformation. Various manifestations of cluster melting will be the subject of section 3. Once the cluster has reached its liquid state, it may evaporate atoms. The issue of cluster stability with respect to dissociation will be addressed in section

4 in relation with the cluster liquid-gas transition. In particular, we will question the extent to which a boiling point can be defined for such small systems.

In the following, our main concern will be to provide the reader with a general flavor of what numerical simulations teach us about the behavior of atomic and molecular clusters when the temperature is varied as the main external parameter. The details of the simulations and the description of the algorithms will be omitted in this overview, but the appropriate references will be given in the bibliography section.

## 2 Solid-Solid Transitions

The solid phase of a cluster is associated with a specific structure, or a limited set of structures, if local defects are allowed. As in the bulk, solid-solid transitions can be induced by varying some external parameter (usually temperature, pressure or density). However, size itself being a crucial variable, clusters often change structure as they grow. This is actually necessary for most materials, which are not crystalline at very low sizes. The typical size where clusters start to adopt the bulk morphology highly depends on the atom or molecule under consideration. This question will not be treated here, since it does not involve temperature or thermal effects directly. A very interesting and recent review on this subject can be found in reference [35].

Archetypal structural transitions driven by temperature occur when two isomers on the potential energy surface (PES) are close in energy and separated by a rather high barrier. However, additional conditions are required for a transition to really occur. For this, and following Bixon and Jortner [36], we consider a very simple 2-state model, where the $N$-atom cluster has two isomers (1) and (2) in thermal equilibrium at temperature $T = 1/k_B\beta$, $k_B$ being the Boltzmann constant. These isomers have potential energy 0 and $\Delta E$, respectively. For simplicity we assume that the PES is harmonic for each isomer, with average geometrical frequencies $\bar{\nu}_1$ and $\bar{\nu}_2$ for isomers (1) and (2), respectively. The total canonical partition function of the system is

$$Z(\beta) = \frac{1}{(\beta h\bar{\nu}_1)^{3N-6}} + \frac{\exp(-\beta\Delta E)}{(\beta h\bar{\nu}_2)^{3N-6}} = Z_1 + Z_2, \tag{1}$$

where $h$ is the Planck constant. The statistical weight of isomer (2) relative to isomer (1) is given by

$$w = \frac{Z_2}{Z_1} = \left(\frac{\bar{\nu}_1}{\bar{\nu}_2}\right)^{3N-6} \exp(-\beta\Delta E). \tag{2}$$

A structural transition from isomer (1) to isomer (2) occurs if $w > 1$, or equivalently if

$$T > T_{\rm ss} = \frac{\Delta E}{(3N-6)k_B \ln \bar{\nu}_1/\bar{\nu}_2}. \tag{3}$$

If $\bar{\nu}_1 \leq \bar{\nu}_2$, then the crossover temperature $T_{\rm ss}$ is not defined, and the transition does not occur. This means that a structural transition to a higher energy structure also needs this structure to have a smaller frequency, that is a broader basin. Fig. 1 illustrates this entropic condition.

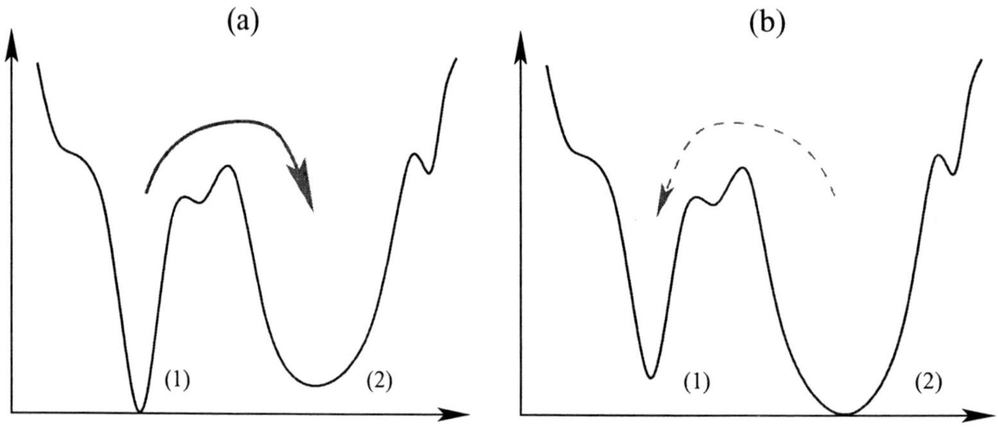


Figure 1: Two schematic, one-dimensional potential energy surfaces likely to exhibit a structural transition induced by temperature. In (a), the basin of isomer (2) is broad and has a higher entropy than (1). In (b), the global minimum also has the highest entropy, no transition occurs.

## 2.1 Lennard-Jones Clusters and Related Systems

The energy landscape of the 38-atom Lennard-Jones cluster is characteristic of a two-funnel system [37], as depicted in Fig. 1(a). The global minimum of this cluster is a truncated octahedron, which is an unusual morphology in this size range [38]. Several icosahedral structures lie slightly higher in energy above the global minimum.

By extending the previously discussed 2-state model to a large set of isomers, one can estimate the partition function of the system assuming harmonic individual partition functions [39], or more accurate empirical [40] or analytical [41] anharmonic perturbative corrections. These so-called superposition approximations predict that a transition between the truncated octahedral structure and the icosahedral isomers takes place close to $T_{\rm ss} \sim 0.1\varepsilon/k_B$, with $\varepsilon$ the LJ unit of energy [42]. For comparison, melting occurs at $T_{\rm melt} \sim 0.165\varepsilon/k_B$ [42].

Conventional Monte Carlo or molecular dynamics simulations are not very practical to simulate the thermal equilibrium between two solid structures, because the large free energy barrier between them takes a very long time (or pseudo MC time) to be crossed [37]. Parallel tempering Monte Carlo [43, 44] is a very convenient improvement that can handle such quasi-ergodicity problems, at least to some extent. In the case of the $LJ_{38}$ cluster, it was possible to obtain quite well converged simulations in both the canonical [45] and microcanonical [46] ensembles. The heat capacity provides a natural quantity for probing the existence of a finite size phase change in the cluster, and exhibits a main peak located near $T_{\rm melt}$. To estimate the relative statistical weight of the octahedral structure, a common method requires performing periodic quenches. Alternatively, the $Q_4$ bond orientational order parameter introduced by Steinhardt and coworkers [47] quantifies the local cubic order, and suitably distinguishes the global minimum from the icosahedral or disordered structures. Fig. 2 shows the canonical heat capacity $C_v$ and the probability of finding the system in its global minimum, in the solid icosahedral structures, or among the other isomers of the liquid state, as a function of temperature.

The simulation results are in semi-quantitative agreement with the predictions of the superposition approximation of Wales and Doye [42]. In particular, the melting temperature and the latent heat of melting (approximately, the surface under the heat capacity peak) are well reproduced. However, even though icosahedral structures indeed occur close to $T \sim 0.1\varepsilon/k_B$, a shoulder in $C_v$ is found at this temperature instead of the small peak reported in Ref. [42].

Icosahedral structures becomes favored over octahedra for entropic reasons. Despite being slightly higher in configurational energy, they have smaller vibrational frequencies, hence a broader funnel in the PES. These "icosahedral precursors" to melting have also been found in metal clusters by the Landman group [48].

More recently, similar phenomena have been identified in clusters of fullerene molecules [49]. Using the simple pairwise Girifalco potential [50] to describe interactions between $C_{60}$ molecules, the $(C_{60})_{14}$ cluster adopts a decahedral-based geometry in its global minimum. Parallel tempering Monte Carlo simulations indicate that this cluster undergoes two transitions upon increasing in temperature [49]. Below the melting temperature (near 800 K), a structural transition occurs, leading to icosahedral minima, which again have only slightly higher energy but a significant vibrational entropy with respect to the decahedral minimum. These results clearly demonstrate that no simple correlation exists between the melting point of a cluster and the properties of the global minimum alone. In general, much larger parts of the energy landscape come into play in the melting process.

Other Lennard-Jones clusters may exhibit a temperature-induced structural transition. Below 30 atoms these clusters have a polyicosahedral or anti-Mackay structure, in which the atoms added to the primitive, 13-atom icosahedron come into capping positions. The 31-atom LJ cluster is the smallest to have a multilayer icosahedral global minimum, obtained by removing outer atoms from the 55-atom icosahedron. Anti-Mackay clusters have lower vibrational frequencies than multilayer icosahedra [51]. This fact is responsible for a structural transition leading to polyicosahedral isomers at a rather low temperature, $T \sim 0.03\varepsilon/k_B$, or about the tenth of the melting point [52]. These results were later confirmed by Frantz who systematically calculated the caloric curves of LJ clusters containing less than 60 atoms [53, 54].

Vibrational frequencies are crucial for the determination of relative configurational entropies, but also for the zero-point energies of stables structures. Quantum delocalization effects and the zero-point motion may be responsible for another type of structural transition [51]. Such effects are of course expected to be most important for weakly bound and light atoms such as neon and especially helium. However a structural transition occurs in the $Ar_{17}$ cluster, whose global minimum deviates from the corresponding LJ minimum after including the ZPE contribution at the harmonic level. Quantum effects can be taken into account by replacing the classical partition functions of Eq. (1) by the corresponding

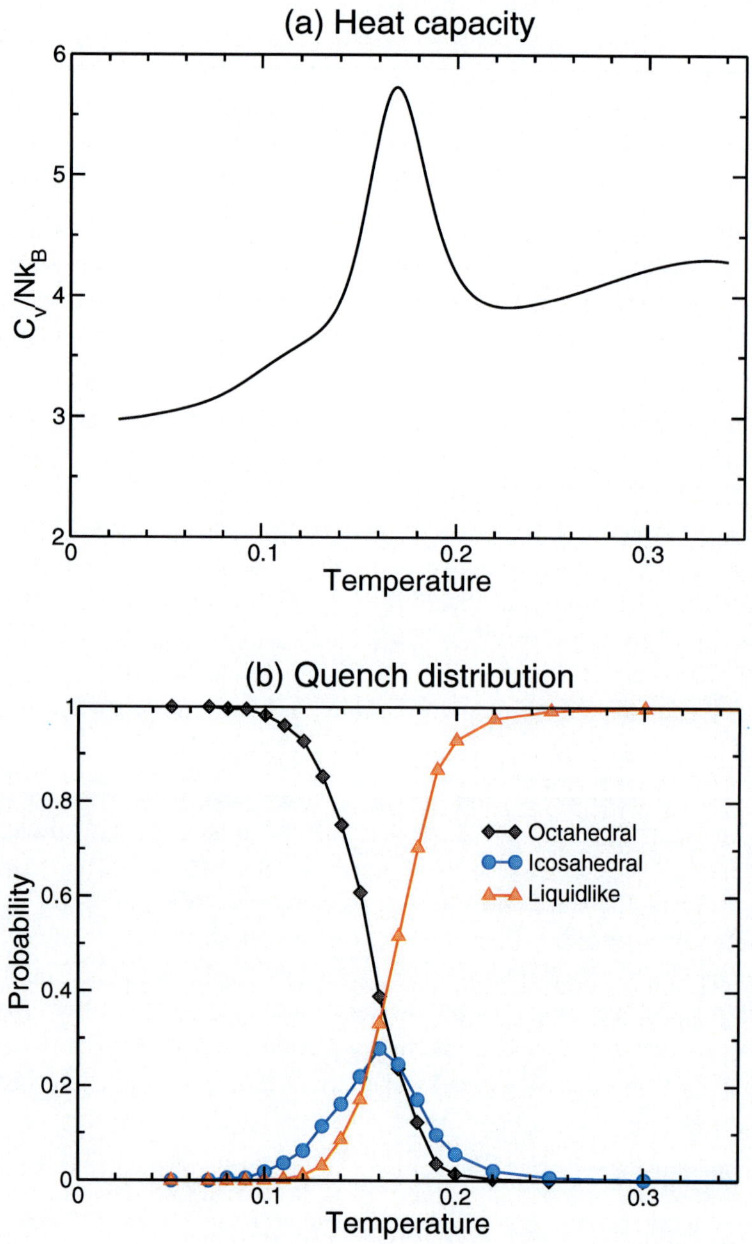


Figure 2: (a) Heat capacity of $LJ_{38}$; (b) probability of finding the system in the octahedral, icosahedral, or liquidlike isomers. The temperature is indicated in reduced LJ units ($\varepsilon/k_B$).

quantum expression for harmonic oscillators:

$$Z(\beta) = \prod_{i=1}^{3N-6} \frac{\exp(-\beta h\nu_{1,i}/2)}{1-\exp(-\beta h\nu_{1,i})} + e^{-\beta\Delta E} \prod_{i=1}^{3N-6} \frac{\exp(-\beta h\nu_{2,i}/2)}{1-\exp(-\beta h\nu_{2,i})}, \tag{4}$$

where the $\nu_{k,i}$ is the $i$th normal mode frequency of isomer $(k)$. Unfortunately, inverting the above equation to find the transition temperature $T_{\rm ss}$ is no longer straightforward.

Calcium-doped argon clusters have a complex melting behavior [55]. In addition to quantum delocalization effects, heterogeneity creates a much richer energy landscape in $CaAr_n$ than in the associated homogeneous $Ar_{n+1}$ cluster [56]. This complexity can be conveniently visualized using the disconnectivity graph representation. After their introduction by Becker and Karplus [57], disconnectivity graphs have been extensively used by Wales and coworkers [58] to represent the most important features of an energy landscape, namely its stationary points and their connectivity. In these graphs, each minimum starts a vertical line at the corresponding energy, and is connected to other minima at the transition state energies. Further details on disconnectivity graphs can be found in the recent book by Wales [59]. Free energy disconnectivity graphs can also be constructed, using harmonic or better [60] approximations for the individual relative free energies of the basins and transition states [56, 61]. Temperature-induced structural transitions are easily identified on such free energy graphs.

In Fig. 3 we have represented the lowest part of the disconnectivity graphs of $CaAr_{37}$ at 0 K and 20 K, along with several low-energy isomers. As can be seen from these graphs, the isomers are rather close in energy, but some are very different in geometry. The decahedral global minimum competes with the truncated octahedron, but icosahedral isomers are not far from these two in energy. Neglecting zero-point effects, the global minimum remains decahedral, but anti-Mackay isomers become disfavored with respect to other icosahedra and the truncated octahedron. Most isomers represented in Fig. 3 see their rank change upon including the ZPE, as was also found in water clusters by Burnham and coworkers [62].

Structural transitions are seen on the free-energy disconnectivity graph at 20 K, where one Mackay icosahedral structure becomes more stable than the decahedral minimum. Unfortunately, the multiple-funnel character of the disconnectivity graph remains and hampers ergodic simulations at low temperature with the presently available computer power, even supplemented with the parallel tempering strategy.

## 2.2 Ionic Clusters

Due to the presence of repulsive forces, alkali halide clusters exhibit fewer isomers than Lennard-Jones clusters with the same number of atoms. Their stable structures are usually based on the cubic crystalline motif, as well as hexagonal or octogonal rings [63]. Isomerization and melting in these ionic systems have been theoretically investigated by several authors [64, 65, 66, 67].

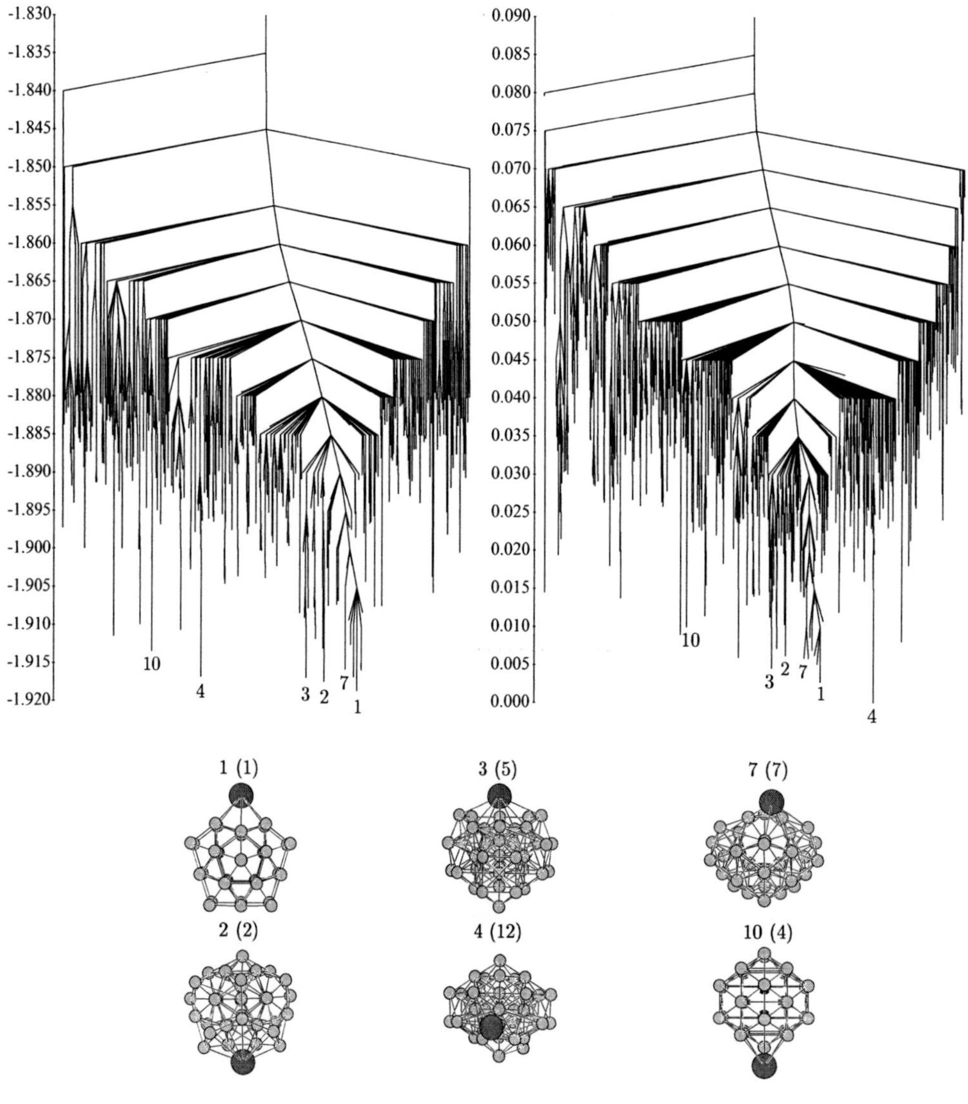


Figure 3: Disconnectivity graphs at 0 K (left panel) and 20 K (right panel) for the $CaAr_{37}$ cluster. Only the lowest parts of the graphs are represented. Several low-energy isomers are represented, with their rank including (or neglecting) the ZPE correction at the harmonic level.

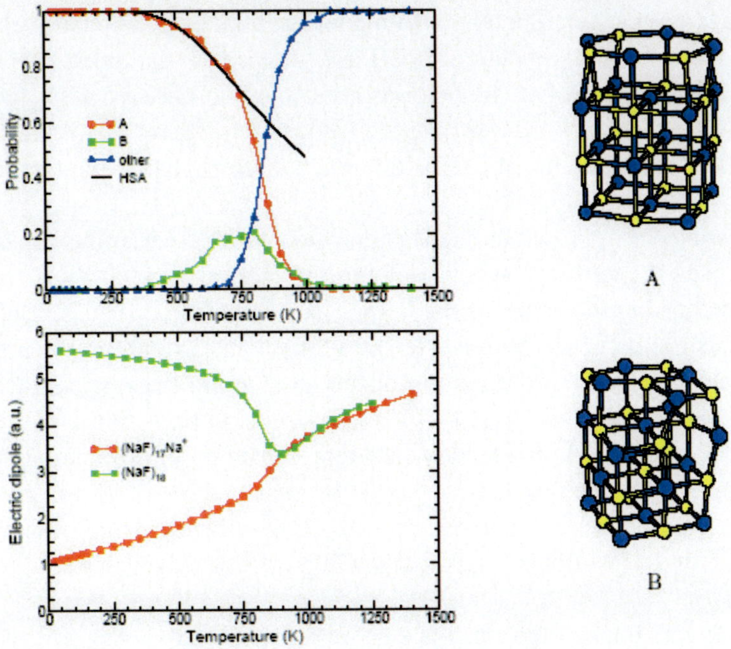


Figure 4: Upper panel: probability of finding each of the two lowest energy isomers of $(NaF)_{17}Na^+$ versus temperature from Monte Carlo simulations (symbols) and the harmonic superposition approach (HSA, solid line). The two isomers A and B are represented on the right side. Lower panel: average electric dipole of $(NaF)_{17}Na^+$ and $(NaF)_{18}$.

Jarrold and coworkers performed ion chromatography measurements on anionic NaCl clusters [24]. In particular, these authors were able to identify different isomers of $(NaCl)_{35}Cl^-$ and to estimate some isomerization rate constants. These experiments were confronted favorably to the theoretical results obtained by Doye and Wales [66, 68], who characterized the rearrangement pathways between the isomers, and employed a superposition method for the thermodynamics and a master equation approach for the kinetics. The time scales obtained for the isomerization processes were in agreement with the measurements [68]. These isomerizations thermodynamically appear as structural transitions [66].

Alkali alide clusters can be considered as nanocrystals, generally based on the $\ell \times m \times n$ cuboidal motif. Rearrangements between isomers based on different sets $(\ell, m, n)$ and $(\ell', m', n')$ are rather cooperative, are tend to involve high energy barriers [66]. Rearrangements between isomers based on the same nanocrystalline motif also exist.

We have illustrated the structural transitions occuring in ionic clusters on the example of $(NaF)_{17}Na^+$. The lowest energy isomer of this cluster is based on the $3 \times 3 \times 4$ cuboid with a missing $F^-$ corner ion. In the second isomer, slightly higher in energy, the missing ion is at the center of a $3 \times 3$ outer face. Other locations of the defects are possible, but the resulting structures are in competition with isomers not based on the $3 \times 3 \times 4$ motif. Parallel tempering Monte Carlo simulations of this system have been performed using a simple Born-Mayer+Coulomb potential to model ionic interactions. Parallel tempering is necessary to overcome the very long time scales for isomerization at low temperatures. Periodic quenches along the MC trajectories allow the two lowest energy structures to be identified among all isomers. The simulation results are given in Fig. 4, and compared to the predictions of the harmonic superposition method, restricted to the two lowest energy isomers.

The simulation and approximation data are consistent with each other, up to some quantitative differences due to the appearance of extra isomers above 700 K. The melting point is indicated by the sharp increase of the higher isomers, close to 800 K. Below this temperature, a structural transition between the two lowest energy minima occurs near 600 K. This transition can be seen as the motion of the missing ion between surface sites. Contrary to what was seen in $(NaCl)_{35}Cl^-$ [66], the structural transition is not evidenced on the caloric curves, and the heat capacity (not represented here) does not show any bump. However, such extra premelting features could be observed for other sizes, which happen to have competing families of cuboids. As will be seen below, larger nanocrystals show sharper melting, and the structural transitions will probably become less and less visible on the caloric curves.

In Fig. 4 the variations of the average electric dipole of the $(NaF)_{17}Na^+$ and $(NaF)_{18}$ clusters with temperature are represented. The latter cluster is the perfect $3 \times 3 \times 4$ cuboid, and has a large dipole in this stable structure. At low temperatures the electric dipole linearly varies, in opposite directions between the two clusters. No special feature is seen in $(NaF)_{17}Na^+$ at the structural transition temperature. However, as both clusters melt near 800 K, the dipole exhibits strong changes before increasing again smoothly above 1000 K. This shows that, expectingly, the two clusters share similar disordered phases. More importantly, it emphasizes the possible use of electric properties to measure melting points in clusters.

## 2.3 Hydrogenated Palladium Clusters

Palladium clusters are of great interest in catalysis because chemical dissociation of organic molecules occurs easily at their contact. They have been investigated theoretically by a number of authors [70, 71, 72], at the *ab initio* or more empirical levels. Bulk palladium has a high absorption capacity of hydrogen, above 50%, but becomes then brittle. Tománek and coworkers [73, 74, 75] developed an empirical many-body potential to study hydrogenated palladium in the bulk and cluster phases by fitting to density functional theory calculations in the local density approximation. This potential was subsequently used to study the influence of hydrogen on melting in palladium clusters [76].

The most stable structures of $Pd_n$ clusters containing up to 60 atoms and modelled using the Tománek potential show a clear tendency toward close-packed structures, except at sizes close to icosahedral shell filling [77]. The absorption sites of hydrogen atoms are tetrahedral, usually located above or below the surface sites. For $Pd_{55}$, the large number of triangular facets yields an absorption capacity higher than 100%. However, at large hydrogen concentrations, the resulting structures poorly resist to temperature jumps [76], and

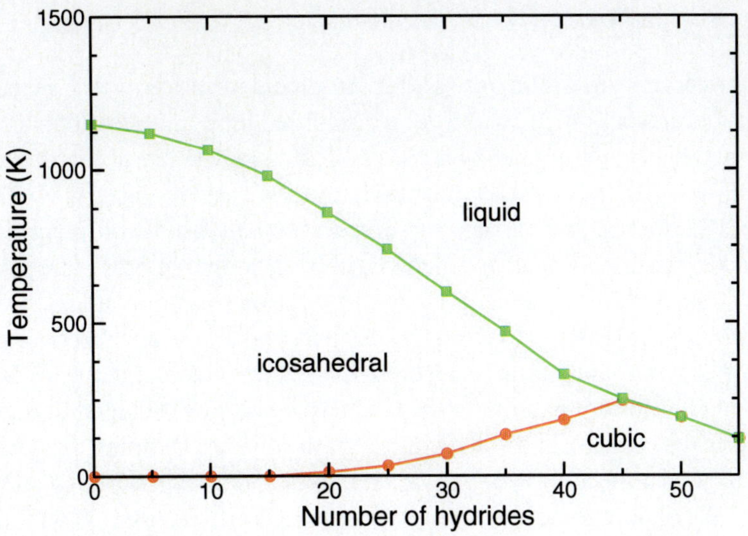


Figure 5: Stability diagram of $Pd_{55}H_m$ clusters in the $(m, \text{temperature})$ plane.

the outer hydrogen atoms tend to recombinate and evaporate as $H_2$. The extra stability of cubic-based structures with a high hydrogen content is due to the better octahedral absorption sites in the fcc material. Hydrogenated palladium is then most stable as in the B1 rocksalt geometry.

Hence, for a given number of Pd atoms, a structural transition is expected as the hydrogen concentration increases. However, icosahedral structures also show slightly lower vibrational frequencies than the cubic structures, and increasing temperature favors icosahedra over cuboctahedra. Fig. 5 shows the $(m, T)$ stability diagram for the $Pd_{55}H_m$ clusters, with $0 \leq m \leq 55$. For each of the icosahedral or cubic states, the lowest free energy structures were obtained by sampling the hydrogen absorption sites and by calculating the corresponding partition functions as the superposition of individual harmonic partition functions. In this diagram the melting points estimated from parallel tempering Monte Carlo simulations have also been reported.

The diagram sketched in Fig. 5 contains information about the size-dependence of the structural transition, which occurs only for $m \geq 20$, and its temperature. A similar picture is found for larger icosahedral clusters, the zero-temperature crossover toward cubic geometries being found at smaller hydrogen concentrations. This should be expected since cubic structures get increasingly stable with respect to icosahedra in free palladium clusters. As seen in Fig. 5 and as noticed previously by Rosén and coworkers [76], hydrogenated palladium clusters have a lower melting point than pure Pd clusters. High hydrogen concentrations also result in higher transition temperatures between cubic and icosahedral structures. Above $m \sim 45$, the range of stability of icosahedral structures vanishes, and melting becomes a single-step process, albeit occuring at low temperatures.

## 2.4 On the Role of Pressure

Nanoparticles interact with a fluid or a solid environment in various ways. Depending on the magnitude of the interactions between the external medium and the atoms of the particle, chemical or metallic bonding may occur, complicated by the possible different facettings of two crystalline systems, or pseudo-crystalline in the case of clusters. Hydrostatic compression forces exerted by the surrounding of a cluster can also be of importance. Clusters of semiconducting materials such as silicon have been shown in brute-force molecular dynamics simulations [78, 79] to undergo an amorphization transition at high pressures.

Beyond the full atomistic modelling of a cluster inside a macroscopic environment, accounting for an external pressure is also possible phenomenologically, by assuming that the system is at equilibrium in terms of its Gibbs free energy. This quantity contains a $PV$ term, where $P$ is the pressure and $V$ the volume of the cluster. Defining a volume for a small number of atoms or molecules is not obvious, however this notion seems correct for large nanoparticles. We choose to measure the volume of a $N$-atom cluster from its convex hull, *i. e.* the smallest convex polyhedron containing all atoms. Standard routines for computing convex hulls are easily available [80]. This approximation is suitable for rigid clusters with very regular and symmetric structures, such as icosahedra or octahedra, but would be poor for clusters having defects or large concave regions, such as the Dzugutov clusters studied by Doye and coworkers [81]. The difference in Gibbs free energy between two structures (1) and (2) is expressed by

$$\Delta\mathcal{H} = \Delta E - (3N-6)k_BT\ln\frac{\bar{\nu}_1}{\bar{\nu}_2} + P\Delta V, \tag{5}$$

with $\Delta E = E_1 - E_2$ and $\Delta V = V_1 - V_2$, $E_i$, $\bar{\nu}_i$ and $V_i$ being the energy, average geometrical frequency and volume of structure $(i)$, respectively. In the above equation, and for sake of simplicity, we have again assumed the cluster to behave harmonically and classically. A finite pressure can slightly affect the most stable configuration of the cluster, because its equilibrium structure is now determined by minimizing $E+PV$ instead of simply $E$. Therefore, $E$, $\bar{\nu}$ and $V$ all strictly depend on $P$. The approximation, which consists in neglecting the pressure-dependence of these properties still provides valuable information about the relative stability of various clusters morphologies at finite temperature and pressure, and actually works quite well [82].

Coarse-grained (temperature,size) stability diagrams at zero [83, 84] or finite [82] pressure are obtained by considering only the properties of "magic" sizes corresponding to filled geometric shells, and by interpolating for intermediate sizes. For this the cohesion energy per atom, the average geometric frequency, and possibly the volume per atom, are fitted as polynomials in $N^{-1/3}$. At a given size, a (temperature,pressure) diagram can also be inferred from the relative Gibbs free energies. We have represented in Fig. 6 such a diagram for selected silver clusters with sizes 670 and 1000, respectively, modelled using the many-body empirical potential with parameters taken from Ref. [85].

At low pressures and low temperatures, both clusters are preferentially found in the cubic shape. In $Ag_{670}$, keeping a low pressure but increasing temperature, two structural

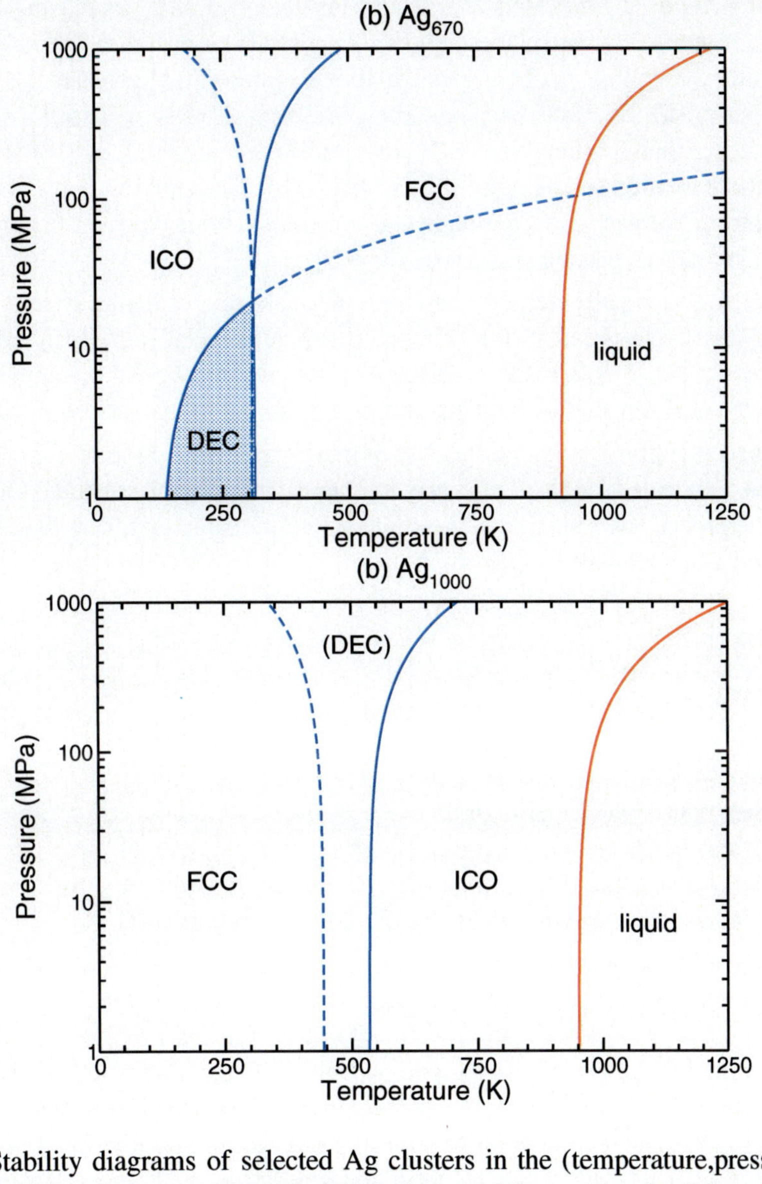


Figure 6: Stability diagrams of selected Ag clusters in the (temperature,pressure) plane. ICO, DEC, FCC refer to icosahedral, decahedral, and face-centred cubic structures, respectively. (a) $Ag_{670}$ shows an intermediate range of stability in the decahedral shape; (b) $Ag_{1000}$ is only stable in the icosahedral and cubic shapes.

transitions lead to decahedra, then icosahedra, which are entropically more and more favored over octahedra. The greater relative stability of cubic, decahedral and icosahedral morphologies has been observed previously and discussed for a variety of atomic clusters, including Lennard-Jones and Morse clusters [83, 84, 86]. Increasing the pressure destabilizes the non-spherical morphologies. This tends to decrease the range of stability of decahedra, which are no longer stable above some critical pressure $P^* \sim 20$ MPa. At even higher pressures, cubic shapes resist more because they sustain the extra internal strain much better than icosahedra, which are naturally strained [82, 86]. This explains the increase in the transition temperature seen in Fig. 6(b). The melting point, estimated and extrapolated from Monte Carlo simulations, also increases with pressure, in agreement with the previous findings by Cheng and coworkers [87].

In $Ag_{1000}$, only one structural transition from cubic to icosahedra is seen, decahedral shapes being less stable already at 0 K. The intermediate range of stability in the icosahedral structure is also reduced with respect to smaller sizes, the icosahedral precursors to melting being estimated to disappear at about $10^4$ atoms within this model, at zero pressure.

The influence of pressure on the stability of more general solid structures can be investigated similarly, provided that a reliable way to quantify the cluster volume is available. The convex hull approach used here could be replaced by the simpler spherical approximation or, alternatively, by alpha hull approximations.

# 3 Melting

In this section we examine clusters of various materials exhibiting the solid-liquid transition, and we focus on the melting process itself in more details. Experimentally a free cluster never reaches thermal equilibrium because evaporation takes place after some long but finite observation time. This limitation is overcome in simulation by constraining the cluster into a container with soft (MD) or rigid (MC or MD) walls [17].

Cluster melting has been the subject of very numerous works, most often using explicit, empirical potentials which have the advantage of being computationally cheap. More recently, similar studies have been undertaken with quantum chemical descriptions, either at the tight-binding [88, 89] or density functional theory [90, 91, 92] levels. However, the case of $LJ_{38}$, presumed simple from its size and interatomic forces, demonstrates that simulations are sometimes very difficult to converge due to the multiple-funnel nature of the energy landscape. Such features are responsible for a large part of cluster size effects, at the same time they cannot be anticipated from the knowledge of the global minimum structure alone.

As far as possible, the following examples have been chosen from simple atomistic models, and the simulations were performed after a careful analysis of the low-temperature structural properties. For each type of clusters, the tools for detecting the onset of melting are described.

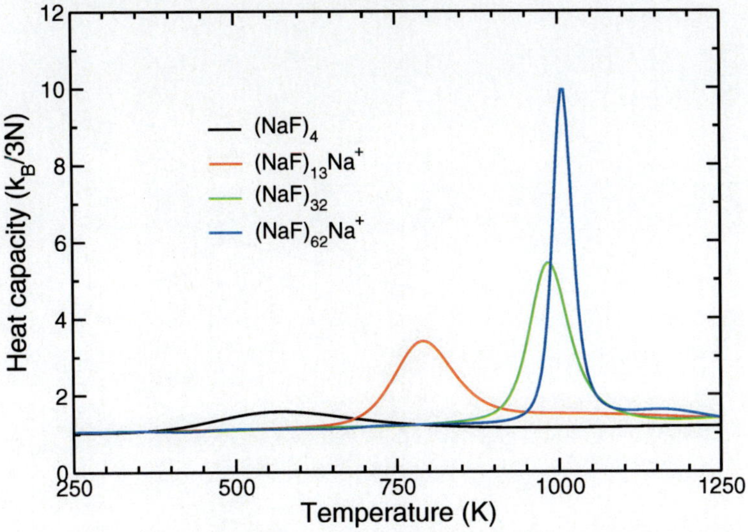


Figure 7: Reduced heat capacities of cubic sodium fluoride clusters.

## 3.1 Ionic Clusters: Size Effects and Dynamical Coexistence

Cubic alkali halide nanocrystals form the magic numbers for such ionic species [63]. These particularly stable clusters do not generally suffer from competing low-energy isomers, and they exhibit a rather clear melting behavior. NaCl clusters have probably been the most studied [64, 65, 93, 94, 95, 96]. Isomerization in the NaCl tetramer was investigated by Heidenreich and Jortner [65]. Berry and coworkers performed molecular dynamics simulations of small [93] and larger [94, 95, 96] clusters, paying a special attention to the dynamical coexistence problem.

Here we focus on sodium fluoride clusters, modelled using the same pairwise potential as in Sec. 2.2. We consider only the $(NaF)_n$ and $(NaF)_pNa^+$ clusters with $n = 4$ $(2\times2\times2)$, $n = 32$ $(4\times4\times4)$, $p = 13$ $(3\times3\times3)$ and $p = 62$ $(5\times5\times5)$, respectively. These clusters are the first two magic numbers for the neutral and cationic series. Their canonical heat capacities $C_v(T)$ are represented in Fig. 7.

For each cluster size, the single peak in $C_v$ is the signature of the melting phase change. The heat capacity is nearly constant at low temperatures, but actually slightly increases due to anharmonicities. Above the melting range, it is again constant but with higher values than in the solid state. The heat capacity peak gets narrower and higher for larger sizes, and tends to the delta function of the bulk material, which characterizes a first-order transition [17]. A convenient definition for the melting point is provided by the central temperature of the peak. As the cluster grows, its melting point is shifted to higher temperatures, but still remains below the bulk melting point. This picture is in agreement with the pioneering work by Buffat and Borel [23] on deposited gold nanoparticles, and confirms simple models that predict the depression in the melting point as $\Delta T \propto 1/R$ [4].

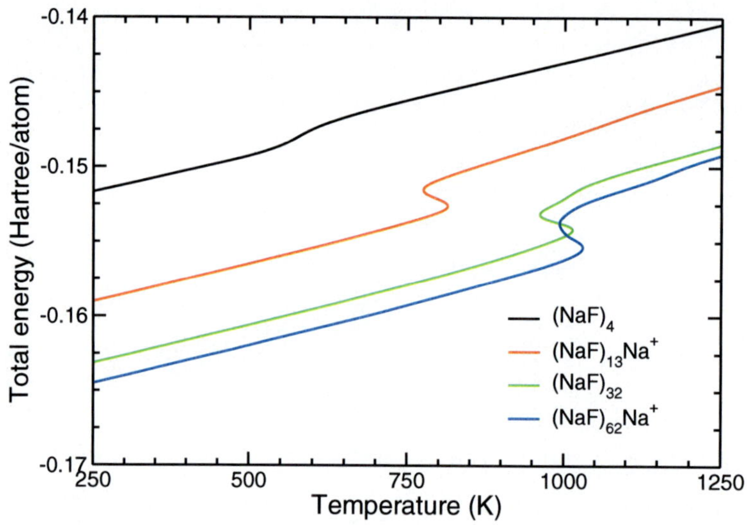


Figure 8: Microcanonical temperature versus total energy in cubic sodium fluoride clusters.

The microcanonical quantity of interest is the kinetic temperature. At least two possible definitions for the microcanonical entropy are available, which lead to different expressions for the temperature and heat capacity [97]. Their differences are negligible in the bulk limit, and remain tiny even for small clusters. Fig. 8 shows the microcanonical caloric curve in the (temperature,total energy) plane for the same sodium fluoride clusters.

The corresponding canonical curves (internal energy versus temperature, not represented) match the microcanonical temperature outside the melting range. However, for the larger clusters, they differ qualitatively and display a S shape (or van der Waals loop) at melting. Such a phenomenon was first observed in Lennard-Jones clusters [17], despite some controversy on $LJ_{13}$ due to a lacking of convergence in early simulation studies.

A van der Waals loop in the kinetic temperature is equivalent to an inflection in the microcanonical entropy (sometimes labelled as a *convex intruder*), a bimodal potential energy distribution, or a negative heat capacity [17, 98]. Free energies versus potential energy as the order parameter also exhibit two minima in the temperature range of coexistence [99]. In the microcanonical ensemble, heat capacities may become negative at finite size, but not in the bulk limit where the canonical ensemble is rigorously recovered. Systems with long-range forces, which do not have a thermodynamic limit, as well as non-ergodic systems can also exhibit negative heat capacities [100]. In the present case, the decrease in the kinetic temperature means that the cluster gets cooler at the onset of melting. The interpretation, due to Berry and coworkers [11], is now generally accepted as the dynamical coexistence between a hot solid and a colder liquid, which have the same total energy but different potential energies.

The relationship between dynamical coexistence and thermodynamical observables has been examined in detail by Labastie and Whetten [17] and Wales and Berry [98]. Bixon and Jortner showed that the simple two-state model with appropriate and realistic parameters exhibits S-bends in the microcanonical caloric curve [36]. A semi-analytical model developed by Wales and Doye [101] suggests that van der Waals loops in the kinetic temperature tend to vanish due the increasing importance of surface melting. A work by Rose and Berry on $(NaCl)_{32}$ [95] showed that this ionic cluster does not exhibit surface melting, but instead a peculiar kind of phase separation called 'nonwetting' by these authors. This process could play a role similar to surface melting in homogeneous clusters, and lead to the same conclusions for larger clusters, as predicted by Wales and Doye.

Other questions related to dynamical coexistence remain presently open. Short-time averages of the instantaneous kinetic temperature have long been used by the Berry group to get a better view of the coexistence phenomenon [11]. However little is known about the time period required to get a bimodal signature characteristic of coexistence. This quantity should depend mainly on the cluster itself, meaning its size and type of atoms, but also on the energy. The residence times in the solid and liquid states have not been studied yet. The shape of their statistical distributions would be useful to address the random character of dynamical coexistence.

## 3.2 Sodium Clusters

The experimental data gathered by Haberland and coworkers and by Jarrold and coworkers on metal clusters strongly called for the attention of theoretical interpretations, and several groups attempted to reproduce these measurements, with a variable success. The surprisingly high melting point in tin [25] or gallium [102] clusters could be interpreted by simulations using quite sophisticated electronic descriptions [103, 104]. In the case of gallium, the covalent nature of chemical bonding (rather than metallic) was argued to explain the stronger stability of the solid state [104].

On the other hand, the large amount of data available on sodium clusters [26, 27, 28, 29, 105] is still poorly understood, despite a significant theoretical work on this subject. This situation is mainly due to the quite large clusters involved in the experiment, containing at least 55 atoms. At such sizes, studies based on electronic structure calculations are very limited, even with the relative chemical simplicity of the sodium element. Simulations based on density functionals have started to appear [92, 107], however they remain limited to short time scales and/or small clusters.

We performed parallel tempering Monte Carlo simulations of neutral and cationic sodium clusters in the size range $8 \leq n \leq 147$, using two different atomistic models [108, 109, 110, 111]. A simple many-body empirical potential was employed to search the most stable configurations of the clusters, through a basin-hopping [38, 112] global optimization scheme. This potential was parameterized by Li and coworkers [114] for sodium and other metals, and was also used by the group of Garzón [113]. Another model is the quantum tight-binding Hamiltonian developed by Poteau and Spiegelmann [115]. This more realistic model accounts for electronic shell effects, and was parameterized on the properties of small clusters, namely the dimer and tetramer [115]. The lowest energy structures of clusters modelled using the tight-binding Hamiltonian were located using a growth algorithm [116] as well as basin-hopping optimization [117].

Preliminar investigations using these models [88, 108, 109] have shown that most clusters display a multiple stage melting process. Multi-step melting occurs on the heat capacity by the presence of one or several humps or peaks before the actual main melting peak takes place. In the case of $Na_{20}$, for instance, a kind of surface melting process was observed [88] in which the outer 7 atoms are able to move rather freely over a rigid 13-atom icosahedron. In some cases, isomerizations without latent heat are also observed.

Interestingly, it was found that the relative magnitude of these "premelting" effects seems to decrease with increasing cluster size. This transition between a molecular regime toward a more macroscopic regime in cluster melting was subsequently confirmed by other studies using different models such as the orbital-free density-functional study of Aguado and coworkers [90]. Since no premelting effect has been observed yet in the experimental results of Schmidt *et al.* on rather large clusters [26, 27], this progressive transition to the bulk behavior was thought to solve this qualitative disagreement. Premelting would then require small sizes to be detected, but then the measurements would have to overcome the weakness of the thermodynamic features.

Another more important difficulty concerns the complex size variations of the melting point and latent heat seen experimentally [27, 105]. There is little doubt that these variations result from the interplay between geometric and electronic shell effects, even though the former have recently been shown to be probably the main cause [106]. It is therefore not surprising that the many-body potential, which lacks a proper electronic description, does not reproduce the measurements. We have represented in Fig. 9 the melting points and latent heats obtained from this model, and compared them with the experimental values. The melting point was estimated from the heat capacity maximum and the latent heat was inferred from the shape of the caloric curve, by fitting the solid and liquid parts to straight lines, and calculating the difference between these lines at the melting temperature. These procedures are identical to the experimental analysis.

While the latent heats are reasonably close to the measurement, the melting points are not reproduced. The experimental maxima at sizes 55 and 142 and the minima at sizes 92 and 120 are not found, but the general order of magnitude is correct. At this point, the empirical many-body potential seems unappropriate to describe sodium clusters in a wide size range. This probably comes from its parameterization, only made on bulk properties. Conversely, the tight-binding quantum model significantly overestimates melting points and latent heats [109], which is related to the parameterization on small clusters only [115]. A possible improvement would be to allow the parameters to include some size dependency, however this would be tedious, in addition to not being neat. The influence of a finite electronic temperature was tested in the tight-binding model, by combining the exchange moves of parallel tempering Monte Carlo with swap moves between electronic states [110]. This did not significantly improve the experimental comparison.

Reexamination of the caloric curves in the complete size range $55 \leq n \leq 147$ recently lead to a surprising result. Clusters containing around 130 atoms exhibit two heat capacity peaks, which evolve continuously for increasing size, as illustrated in Fig. 10.

The low-temperature peak progressively grows and shifts to the main melting peak. In

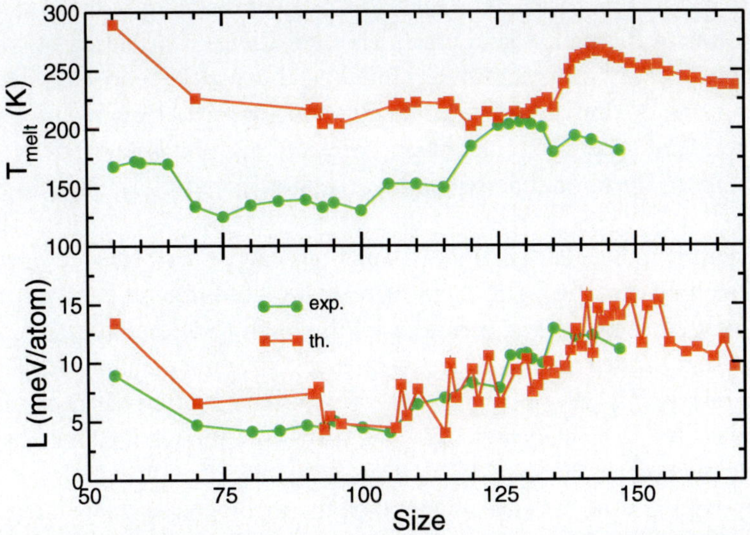


Figure 9: Melting point and latent heat of melting in sodium clusters modelled using an empirical many-body potential. The simulation data are compared to the experimental values. Adapted from reference [111].

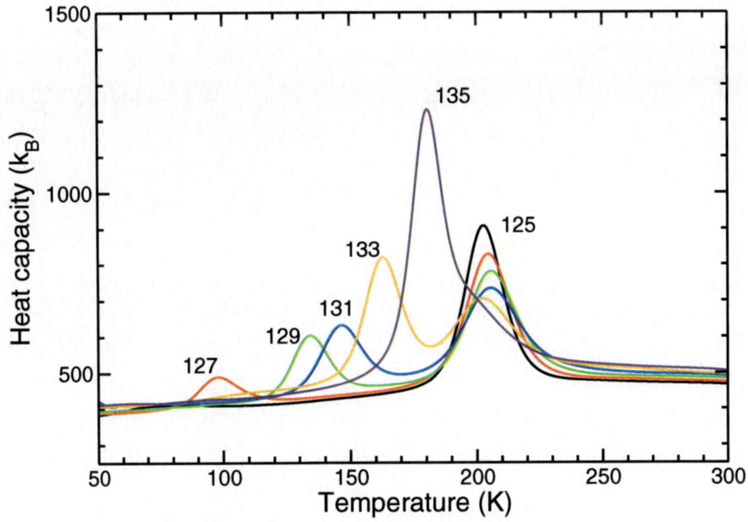


Figure 10: Canonical heat capacities of the clusters $Na_n$, $125 \leq n \leq 135$, described using a many-body empirical potential.

$Na_{133}$, the premelting peak is even higher than the melting peak itself; this case could be referred to as 'post-melting.' Periodic quenches help in elucidating the cause of this phenomenon in terms of the energy landscape. The probability distribution of configurational energies of more than $10^4$ different isomers of $Na_{133}$ is actually found to be clearly bimodal, the two peaks being of similar width and height, and separated by a significant energy gap [111]. The correlation between the energy spectrum and the heat capacity confirms the landscape picture of cluster thermodynamics, and offers here a clear exemple of premelting in clusters.

Complementary molecular dynamics simulations can be performed to interpret dynamically the premelting features. The rapid increase in the diffusion constant indeed shows that the first heat capacity peak coincides with the melting of the outer, uncomplete shell of the cluster. This surface melting process starts at higher temperatures for larger clusters, because the outer shell is more bound to the core. Surface melting has been often observed in simulations of atomic clusters [88, 118, 119, 120], even though its thermal signature was not always established.

As the premelting peak becomes higher than the second peak, the melting point undergoes a strong drop, simply due to the way it is defined. This may shed light on processes, which might be responsible for the complex variations in the experimental melting point, in addition to the shell filling effects alone.

The multiple structure of the caloric curves for sodium clusters having more than 130 atoms is *qualitatively* distinct from the experimental measurement. This is another indication of the unappropriateness of the many-body potential. Premelting features are also found when using the tight-binding quantum Hamiltonian, however they are generally less pronounced than the results obtained with the classical potential [109]. This is true for clusters with about 130 atoms, where preliminar peaks still exist without competing with the main melting peak [111]. Before fully correlated calculations on large clusters with high statistics become feasible, such simple quantum models probably provide a suitable description of the thermodynamics of sodium clusters — possibly after some new parameterization.

## 3.3 Calcium-Doped Argon Clusters

Spectroscopy is an unvaluable tool to probe cluster structure and dynamics. Since the seminal experimental work by Hahn and Whetten [18], several groups have attempted to use spectroscopic measurements to detect a possible signature of isomerization or melting [19, 20]. Temperature effects on the spectroscopic observables are twofold. Firstly, the spectral lines are broadened and shifted due to the vibrational motion. Secondly, the presence of extra isomers can induce new peaks to appear. As noted by Haberland and coworkers on sodium clusters [20], these lines are generally difficult to detect, precisely because of the broadening, especially at small sizes. A related important problem is that, at high temperatures, the thermal shift hampers the attribution of the lines to specific isomers. Furthermore, isomerizations in small clusters take place in broad temperature ranges, hence the spectroscopic signals vary very smoothly. This explains the relative failure of the direct spectroscopic method to measure melting points in small clusters, both experimentally [20] and theoretically [121]. Yet it should be noted that a recent theoretical work by the Landman group [107] was able to relate the variations in the absorption spectrum of some $Na_n^+$ clusters, $3 \leq n \leq 9$, to the atomistic motion and the details of the isomers.

Chromophore-doped rare-gas clusters offer good candidates to look at the influence of phase changes on the spectroscopic properties. The spectroscopy of the chromophore atom or molecule is usually very sensitive to its local coordination, hence it is hoped that the spectral lines could exhibit some qualitative variation upon isomerization, provided that the different isomers actively involve the chromophore. The work by Curotto and coworkers on argon clusters enriched with the HF molecule actually showed that solid-solid transitions and melting could be evidenced using vibrational spectroscopy of the chromophore [122, 123, 124].

The case of calcium-doped argon clusters is interesting from the experimental point of view, because measurements on large clusters are available from the Mestdagh group [125]. In these experiments, the rare-gas cluster is used as a substrate for a chemical reaction between two species, monitored through optical spectroscopy. The cold temperature of argon clusters, close to a few tens of Kelvin, also allows a better resolution of the spectral signals with respect to the same reactions in gas phase. A review paper on these so-called cluster isolated chemical reactions is given in Ref. [126].

Our study of $CaAr_n$ clusters [55, 56, 125, 127] relies on a simple quantum model for the ground and first excited states surfaces. The details of this model, based on the diatomics-in-molecule (DIM) approximation, are given elsewhere [125]. To account for quantum vibrational effects, we use the superposition approximation with harmonic quantum partition functions, corrected perturbatively at first order. The quantum superposition approximation has been shown to predict quantitatively correct thermodynamical properties with respect to path integral Monte Carlo simulations [51]. For comparison, we also carry out a Monte Carlo calculation using the Feynman-Hibbs quantum corrections [128].

The absorption spectrum of the clusters is calculated using the Condon vertical approximation and the semi-analytical theory developed by Wadi and Pollak [129]. In this approach, each absorption line is assumed to be Gaussian, and the position and width of the line explicitly depend on temperature and the local characteristics of the potential energy surface of the ground and excited states. This approach also includes quantum vibrational effects and is naturally suited to work with the quantum superposition approximation. Further discussion on these methods can be found in Ref. [55].

The heat capacities and absorption spectra for the $CaAr_{13}$ cluster are represented in Fig. 11. The role of quantum delocalization is clearly visible at low temperatures, in agreement with our study on pure rare-gas clusters [51]. In its most stable structure, $CaAr_{13}$ is an argon icosahedron capped by the calcium atom. The next stable isomers are based on the same motif, with calcium occupying one of the icosahedral sites and argon as the capping atom. A structural transition takes place around 20 K, without any significant thermal signature in the heat capacity. However, this phenomenon clearly induces changes on the absorption spectrum. Melting is indicated from the maximum of the heat capacity, but does

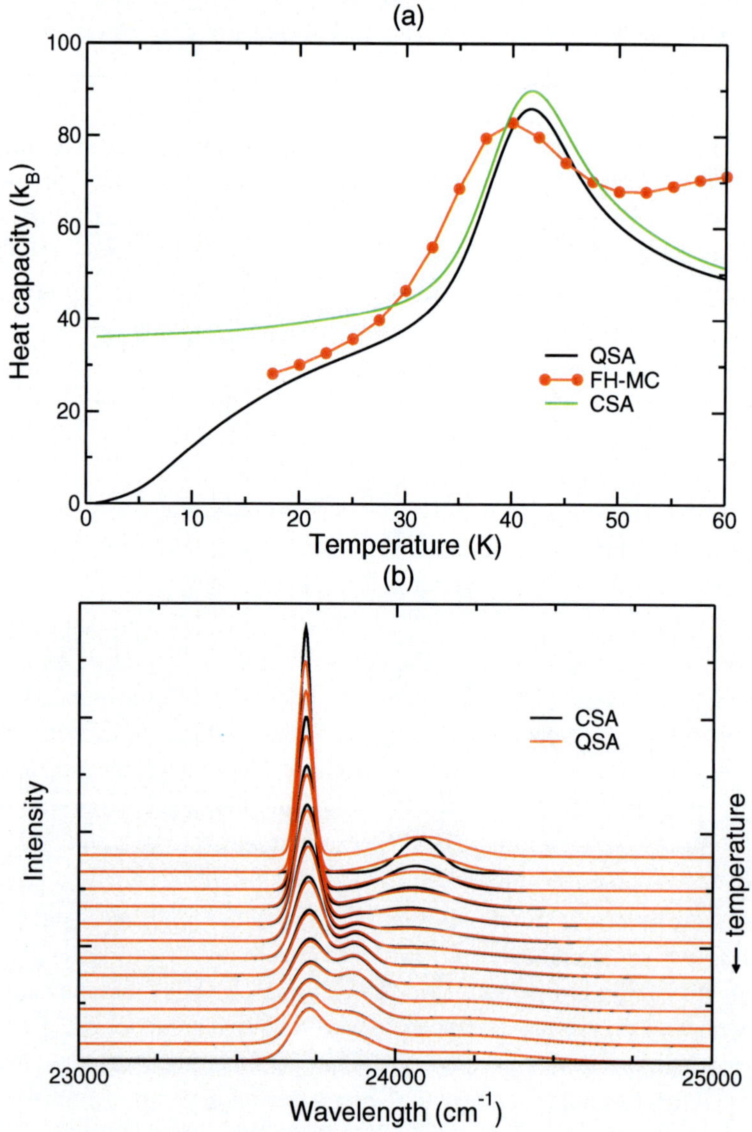


Figure 11: (a) Canonical heat capacities of $CaAr_{13}$, calculated from quantum-corrected Monte Carlo simulations (FH-MC), and from the quantum (QSA) and classical (CSA) superposition approximations; (b) absorption spectra in the quantum (QSA) and classical (CSA) superposition approximations. The spectra are represented for temperatures in the range $0 \leq T \leq 60$ K, every 5 K.

not lead to spectroscopic variations.

The global minima of other $CaAr_n$ clusters containing up to 150 atoms are usually based on the icosahedral geometry, the calcium atom occupying an external site. Calcium being less bound to argon than argon itself [125], its mobility increases at temperatures below the melting point. In such clusters, the changes in the absorption spectrum are more the signature of surface melting than the actual cluster melting.

As aforementioned, $CaAr_{37}$ is a special case exhibiting several stable geometries and structural transitions at low temperature. These isomers belong to different funnels, and display very different spectral lines. The heat capacities and absorption spectra of this cluster are represented in Fig. 12. In this cluster, quantum effects blur the thermal features of the structural transitions at low temperatures. Coexistence between several different structures results in a very resolved absorption spectrum, and a clear identification of any transition temperature is not possible from the spectra.

However, because the structures belong to funnels separated by large barriers, the relaxation dynamics to equilibrium is expected to be very slow and the cluster may get trapped into one funnel during the experimental (or numerical) time scale. This suggests a possible way to use spectroscopy as a probe of slow relaxation dynamics in clusters, using trapped clusters. Examples of absorption spectra followed in time, and obtained from master equation dynamics, are discussed in Ref. [56].

## 3.4 Molecular Clusters and the Plastic Transition

Molecular materials usually display a much richer thermal behavior than atomic species, as well as multiple solid structures. Water, for instance, is known to exist into about ten different crystallographic phases under high pressures, and at least two supercooled liquid phases separated by a first-order transition have been found in numerical simulations [130].

The temperature-dependent properties of molecular clusters have been investigated for a variety of species, including sulphur hexafluoride [131], water [132, 133], benzene [134, 135], carbon dioxide and nitrogen [136, 137], and acetonitrile [138], among many others.

The atoms in the molecules are generally much more strongly bound to each other than molecules in the cluster, unless the external pressure or temperature is so extreme that it induces chemical reactions. Therefore, at low or moderate temperatures, the molecules are nearly rigid. Note also that the classical picture should be considered cautiously when dealing with light atoms. In water clusters, the zero-point contribution plays an important role in the determination of the equilibrium properties [62].

We have chosen to illustrate the case of the molecular cluster $(O_2)_{13}$, using Monte Carlo simulations and treating the oxygen molecules as rigid. Interaction between oxygen molecules is not as simple as, e.g. between nitrogen molecules, because the $O_2$ molecules can be found in their singlet or triplet state. However a few potentials are available for use in numerical simulations. We compare here the predictions of the Klein-Levesque-Weiss (KLW) potential [139] with those of a more refined model due to Etters *et al.* [140]. The KLW potential, which only assumes repulsion-dispersion forces with the Lennard-Jones form between oxygen atoms of different molecules, was designed to model the $\gamma$ phase

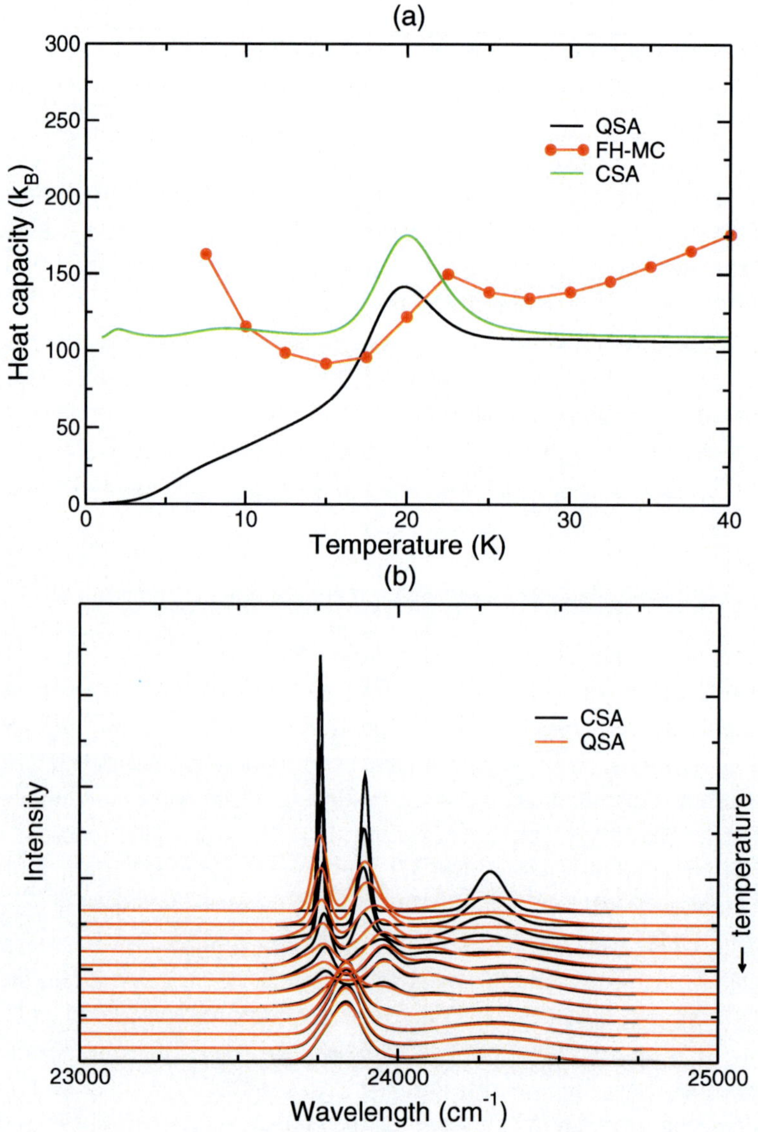


Figure 12: (a) Canonical heat capacities and (b) absorption spectra of the $CaAr_{37}$ cluster. In (b) the spectra are represented for temperatures in the range $0 \leq T \leq 30$ K, every 2.5 K.

of bulk $O_2$. Two sets of LJ parameters were proposed, fitted on the molecular or bulk properties of oxygen, respectively. The Etters potential has a better repulsion-dispersion form with extra $1/r^8$ and $1/r^{10}$ terms. It includes an electrostatic contribution between quadrupoles described using partial charges located along the molecular axis, as well as a spin-orbit term and a Heisenberg Hamiltonian between classical spins. It was designed to reproduce the $\alpha$ and $\beta$ phases of bulk oxygen.

For all models, the lowest-energy minimum of the $(O_2)_{13}$ cluster consists of an icosahedron formed by the molecular centres of mass. The heat capacities are represented in Fig. 13.

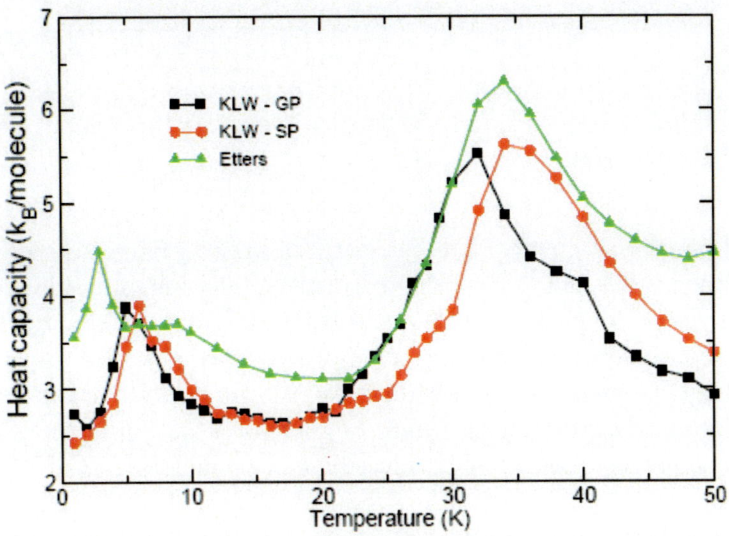


Figure 13: Canonical heat capacities of the $(O_2)_{13}$ cluster calculated using the KLW potential with gas-phase (GP) or solid-phase (SP) parameters, and the Etters potential.

The potentials generally agree with each other. The heat capacity is $3k_B/2$ higher with the Etters potential, due to the extra magnetic degrees of freedom. The two parameterizations of the KLW potential yield similar results, only differing by less than 2 K near the main heat capacity peak.

Interpreting the different peaks in $C_v$ requires appropriate tools. The root mean square bond length fluctuation index $\delta$ conveniently probes the solid-liquid phase change, and generally (but not always [109]) provides a rather accurate determination of the melting point. This parameter is calculated from the first two moments of the distances between each pair of molecular centres of mass. One can similarly define an index for rotational melting of clusters of linear molecules, from the fluctuation of the angle between molecular axes. The variations of these two order parameters versus temperature are shown in Fig. 14 for the KLW potential with gas-phase parameters.

The sharp rise of each parameter marks the onset of a larger mobility of the corresponding degrees of freedom. At low temperature $T \simeq 5$ K, the first heat capacity peak is

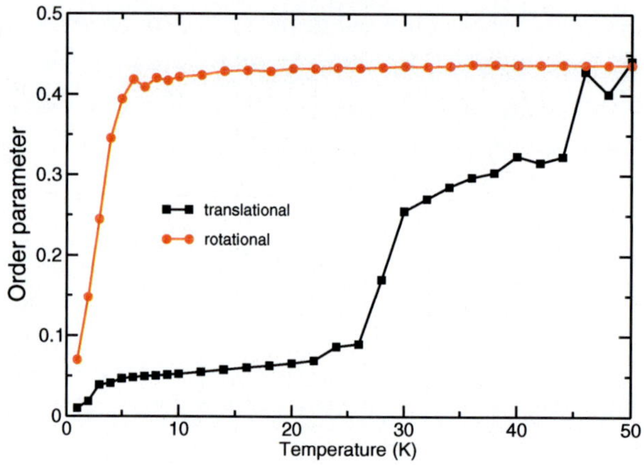


Figure 14: Translational and rotational fluctuation indices in the $(O_2)_{13}$ cluster modelled with the KLW-GP potential.

associated with rotational melting, but the centres of mass of the molecules remain rigid. This phase change is the finite size analog of a bulk solid-plastic transition. At the maximum of the heat capacity peak, $T \simeq 30$ K, the plastic phase is replaced by the true liquid phase through translational melting. The solid-plastic transition can be seen as a particular type of solid-solid phase change, where all the isomers involved are based on the same geometric structure of the molecular centres of mass, only with different relative orientations. This phenomenon is not specific to clusters of linear molecules, and also occurs, e.g., in methane clusters [141]. The behavior of the 13-molecule $O_2$ cluster described with the Etters potential is slightly more complex due to the magnetic interaction. The rather sharp peak seen close to 3 K is indeed due to the release of the magnetic order in the cluster. The solid-plastic transition occurs near 7 K, and is visible as a broader bump in the heat capacity.

Not all molecular clusters exhibit a plastic transition. Clusters of carbon dioxide molecules, for instance, have structural similarities with nitrogen clusters, but all their rotational and translational degrees of freedom melt simultaneously [137]. The difference essentially comes from the larger size of the $CO_2$ molecule, which is closer to the equilibrium distance between two molecules than in $N_2$ or $O_2$. Smaller molecules rotate more easily, and this property further increases the stability range of the plastic phase. Finally, hints of the more complex bulk phase diagram and new kinds of structural changes may start to appear in larger clusters. Specific order parameters should then be introduced to characterize the various crystallographic sites and the mobility of the associated degrees of freedom.

# 4 Toward the Liquid-Gas Transition

Free clusters with enough internal energy dissociate one or several atoms or molecules, and cool down after dissociation. If the energy is added gradually (*i. e.* slowly) from the solid state, dissociation occurs through evaporation for most materials. Evaporation, or thermal dissociation, may be a multiple and sequential process. It may be caused by an initially hot thermalization, by vibrational excitation and/or electronic excitation followed by internal conversion. If the internal energy is very high and deposited briefly with respect to the typical vibrational time, numerous dissociations can occur nearly simultaneously. One then speaks of multifragmentation.

To some extent, the observation time plays in free clusters a role similar to the excitation energy, because dissociation is fundamentally an out-of-equilibrium process. Thus it seems difficult to investigate the vapor phase of a cluster without properly including the time variable, even though attempts to extend conventional simulations to the gas phase by increasing the container size have been reported [142, 143]. Actually, simple thermodynamical arguments show that the final state of a free cluster would be that of several unconnected fragments, which have no longer any excess energy to dissociate further. For clusters such as $LJ_{13}$, this implies that the liquid state would never be reached in simulations due to several prior evaporations.

Accurate descriptions of the evaporation process are available from statistical rate theories [144, 145, 146]. Such models link the reaction rate, kinetic energy released (KER) after dissociation and related quantities to the statistical properties of the reactants and products *at thermal equilibrium*. Therefore, and as first shown by Weerasinghe and Amar [147], information about the thermodynamical states of the reactants and products could be obtained from looking at their dissociation.

Rate theories are not relevant for clusters undergoing sudden multifragmentation, but a few investigations from simplified models and numerical simulations [148, 149, 150] have attempted to find some signature of critical phenomena during the fragmentation of highly excited finite systems. In the context of nuclear physics, these studies used the fragment distribution as the main observable, as well as support from percolation [149], lattice-gas [150] or the Fisher [151] models [148]. They are generally motivated by the search for a possible equation of state of nuclear matter, but it still unclear how they could be adapted to the case of free atomic clusters. Recently, the Gross multifragmentation model [152] has been used to interpret an the experimental fragment distribution in collision-induced fragmentation of small carbon clusters [153].

The liquid-gas transition is prototypical of a first-order phase transition in the bulk, and only becomes second-order at the critical point. To our knowledge, only the works by the Haberland [29] and Bréchignac [31] groups have reported quantitative experimental evidence of this phenomenon by measuring the complete caloric across the liquid-vapor phase of sodium and strontium clusters, respectively. Briefly, what they observed was the stagnation of the final cluster temperature upon increasing its energy. This indicates that more and more atoms dissociate from the cluster as the excitation energy increases, keeping the temperature approximately constant, in agreement with Klots' evaporative ensemble. Noteworthy, a rather similar behavior was found in the work by Strachan and Dorso [154] on a simulated Lennard-Jones cluster.

### 4.1 Evaporation

It is not our goal here to discuss the statistical rate theories commonly employed in cluster physics. A nice review has been written by Jarrold [146]. Rate theories are ubiquitous in experimental studies on free clusters, because they provide information about binding energies and lifetimes, two very fundamental properties. The Rice-Ramsperger-Kassel (RRK) theory [155] and, to a lesser extent the Weisskopf theory [156] subsequently used by Engelking [157] contain relatively few free parameters, and thus are the most frequently used. The so-called phase space theory (PST) developed by a number of authors [158, 159, 160, 161] improves over the former models by a rigorous treatment of angular momentum conservation. PST can also be considered as a version of transition state theory where the loose transition state approximation is used. In other terms, the products *are* the transition state in PST.

The aforementioned theories have been compared to the results of numerically exact molecular dynamics simulations on small Lennard-Jones clusters [147, 162, 163], and more recently on molecular clusters [164]. It was shown by Weerasinghe and Amar that only PST was able to reproduce the details of the kinetic energy released distributions. Moreover, the predicted evaporation rates were seen to be close to the measured rates within a factor less than 3, whereas RRK or Weisskopf theories often underestimate or overestimate the rates by more than 3 orders of magnitude.

Within phase space theory the evaporation rate is expressed as [146, 147, 161]

$$k(E) \propto \frac{1}{\Omega(E)} \int_0^{E-\Delta E} \Omega'(E-\Delta E-\varepsilon)\Gamma(\varepsilon)d\varepsilon, \tag{6}$$

where $\Omega$ and $\Omega'$ are the densities of states (DOS's) of the product and parents, respectively, $\Gamma$ the rotational density of states, $\Delta E$ the binding energy difference between the product and parents, and $\varepsilon$ the total kinetic energy released. Similarly, the KER probability distribution $p(\varepsilon, E)$ is

$$p(\varepsilon, E) \propto \Omega'(E-\Delta E-\varepsilon)\varepsilon\Gamma(\varepsilon), \tag{7}$$

and thus involves the density of states of the *product* clusters only. The thermodynamical properties of the products are fully contained in the DOS, hence quantities derived from DOS, including the KER distribution and the internal energy, shuld share similar features. This is precisely one of the results first obtained by Weerasinghe and Amar [147]. Fig. 15 shows the average KER versus excitation energy upon evaporation of the molecular clusters $(CH_4)_{14}$ and $(N_2)_{14}$. As was found in the dissociation of $LJ_n$ clusters [147], the average KER monotonically increases but may also exhibit two distinct linear regions separated by an inflection point. This behavior is observed in Fig. 15(b), reflecting the existence of a convex part in the microcanonical density of states of the $(N_2)_{13}$ cluster.

Such a phenomenon not only occurs at energies close to the melting point. For instance, the mixed cluster $KrXe_{12}$ undergoes a structural transition between two icosahedral-based isomers, the krypton atom occupying either an outer site (global minimum) or the icosahedral core (second minimum). In the evaporation of a xenon atom from $KrXe_{13}$, the average

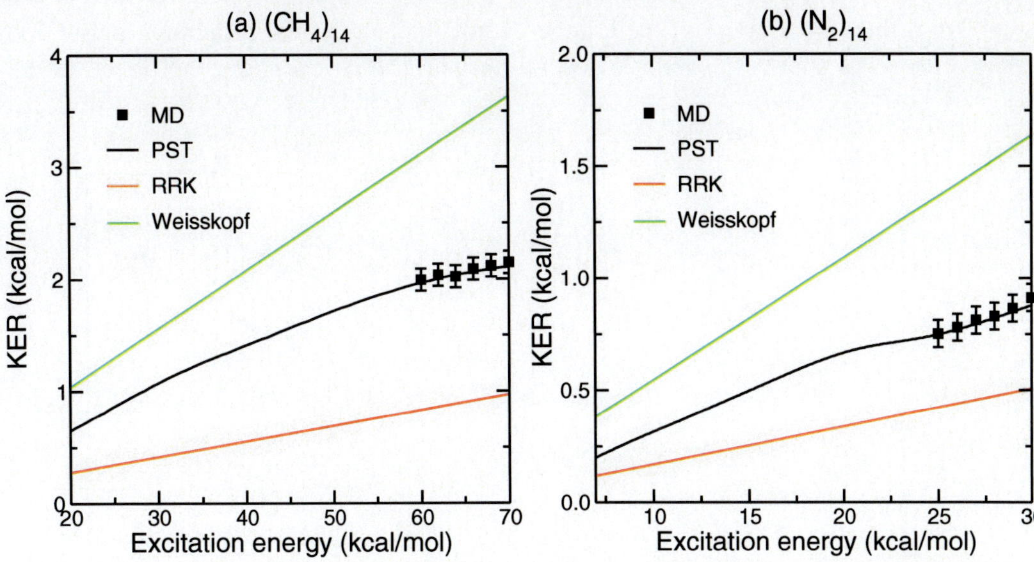


Figure 15: Average kinetic energy released versus excitation energy in (a) $(CH_4)_{14}$ and (b) $(N_2)_{14}$. In both cases, the simulation data (MD) averaged over 5000 trajectories are also reported.

KER clearly shows a signature of this structural transition, in addition to the inflection at melting [165].

The relation between evaporation observables and thermodynamics is thus quite general, and could provide an alternate experimental way of determining melting points in clusters.

## 4.2 Boiling

Thermal relaxation of a highly energetic free cluster leads to a state where a main product is found next to multiple small fragments, usually single atoms or molecules and sometimes dimers [166]. This state highly depends on the observation time. The final temperature of the main product is much lower than the temperature of the initial parent, due to the kinetic energy carried away with each evaporating atom. Because the number of dissociated atoms increases along with the initial excitation energy, the final temperature is approximately constant, if the observation time remains fixed. This is one of the main grounds of Klots' evaporative ensemble [34]. The results from the Haberland group [29] clearly confirm these ideas and show that sodium clusters cannot be heated above some temperature without emitting atoms during the experimental time. This maximal temperature was identified as the boiling temperature.

Classical, isoenergetic molecular dynamics simulations of free Lennard-Jones clusters were performed by Strachan and Dorso across the solid-liquid phase change and up to the multiple evaporation regime [154]. The microcanonical temperature was calculated as usual, from the time average of the kinetic energy but on the main fragment only. The caloric curve was shown to exhibit a second plateau at high energies, interpreted as the

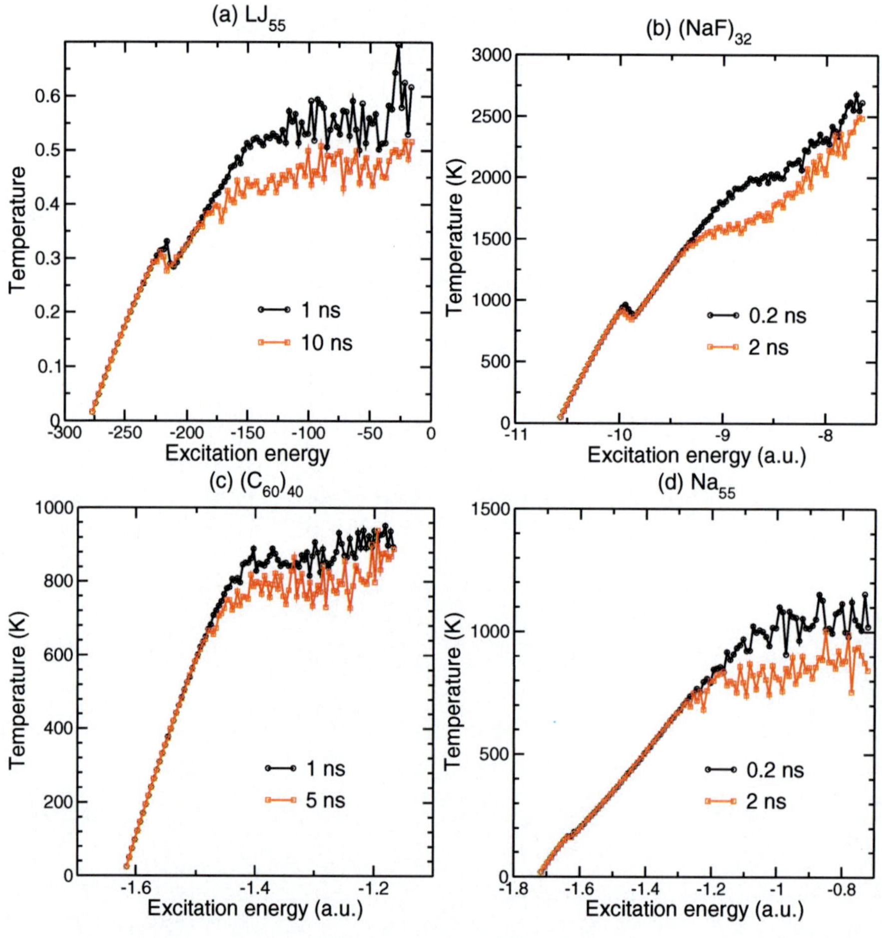


Figure 16: Microcanonical temperature calculated from molecular dynamics simulations and averaged on the main fragment, for different observation times. (a) $LJ_{55}$; (b) $(NaF)_{32}$; (c) $(C_{60})_{40}$; (d) $Na_{55}$.

signature of the liquid-vapor transition. We have carried out similar calculations for various clusters, namely $LJ_{55}$, $Na_{55}$, $(NaF)_{32}$, and $(C_{60})_{40}$. In order to achieve substantial statistical sampling, we use simple explicit potentials to model the clusters. For sodium clusters, sodium fluoride clusters, and fullerene clusters, we use the same atomistic models as aforementioned, namely the empirical many-body, Born-Mayer+Coulomb, and Girifalco potentials, respectively.

The microcanonical caloric curves of these four clusters are plotted in Fig. 16 for two different simulation times. The main fragment is defined as the largest set of connected atoms or molecules, two atoms being connected when their distance is smaller than twice their equilibrium distance. For the Lennard-Jones, sodium, and sodium fluoride clusters, the microcanonical temperature shows a small convex part at low or moderate energies, characteristic of melting. The boiling plateau is seen for all clusters at higher energies. The liquid-vapor transition temperature, as estimated from this plateau, decreases for longer simulation times. Monte Carlo simulations indicate that the fullerene cluster melts in the 800–900 K range [143]. The boiling temperature thus seems to be very close to the melting temperature in this system, which reflects the reported unstability of the liquid phase in bulk $C_{60}$ [167]. In the case of $(NaF)_{32}$, the boiling plateau is quite narrow, and the kinetic temperature starts increasing again above $-8.5$ Hartree/atom. This system can be heated more than the others because internal excitation energy may also be stored in the dissociating NaF molecules. We expect that a more realistic atomistic description of the $C_{60}$ molecules would lead to a similar result, hence a picture somewhat different from Fig. 16(c).

The curves represented in Fig. 16 are quite smooth in the liquid-gas range, and do not allow a precise determination of the boiling energy. The average size of the main fragment, shown in Fig. 17 for the four clusters, provides further information about the fragmentation process. This quantity is expected to be sensitive only to the gas phase, and seems to show a rather similar behavior for all clusters. Above some energy, it decreases linearly, the slope being larger for longer observation times. Only for $(NaF)_{32}$ it reaches a small constant value, but we know that the high energy range is dominated by stronger forces (between ions) than those responsible for evaporation (between dipoles). This critical energy only weakly depends on the observation time, hence it could be used to define unambiguously the boiling point. Comparison with experiments would be more convenient in the canonical ensemble, however it is not clear whether constant temperature molecular dynamics methods — whatever their type — correctly describe fragmentation in thermostatted finite systems.

Contrary to the melting problem, the landscape approach is of little use to understand the liquid-gas transition, which mainly rises dynamical issues independent of the structural details. In this respect, statistical rate theories are more appropriate, but they only describe single dissociation events. Assuming that the multiple evaporations occur sequentially, a semi-analytical calculation of the average fragment size after some finite observation time can be carried out in a fashion similar to the master equation approach used in relaxation dynamics on a potential energy surface. Both techniques may also be combined to predict the relative weights of fragmentation channels involving different sizes or different isomers.

# 5 Conclusion

The effects of temperature on the structural and phase stability of free clusters were considered from the point of view of simulations, with a special emphasis at the relation between phase changes and the energy landscape. Such an approach offers a convenient framework to the theoretical interpretation of irregular melting points as they were discovered in sodium clusters by the Haberland group. It also explains many of the premelting effects seen numerically, including surface melting and structural transitions. Complex cluster size

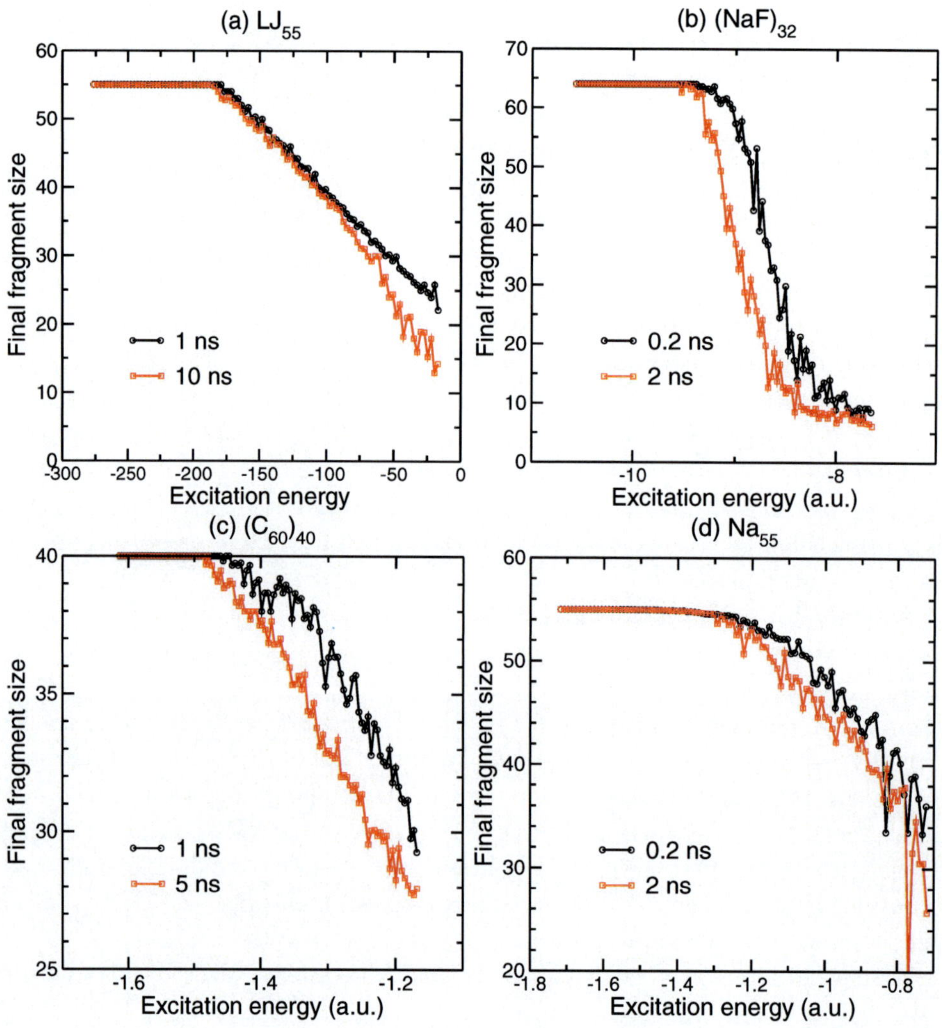


Figure 17: Average size of the main fragment, estimated from molecular dynamics simulations after different observation times. (a) $LJ_{55}$; (b) $(NaF)_{32}$; (c) $(C_{60})_{40}$; (d) $Na_{55}$.

effects make a general theory of melting hardly expected in finite systems, even more than in the bulk where it is also lacking still today.

We have tried to extract as much information as possible from the caloric curve and heat capacity. However, these thermodynamical observables are often difficult to obtain experimentally, and they are not guaranteed to show systematic signatures of the structural changes. In numerical simulations, a variety of indicators such as the Lindemann index are available to trace back the causes for the observed thermal behavior. Melting could be detected using alternate observables, which show some temperature dependence through their dependence on atomic structure. In the present review, the absorption spectrum of chromophore-doped argon cluster was discussed. Similarly, the reactivity of clusters, as suggested by Schmidt and Gerber [168], should also be sensitive to the location of the reactive complex in the host cluster. However, and to our knowledge, reactive collisions have not been used in the context of cluster thermodynamics yet. Electron diffraction patterns were used by Martin and coworkers to estimate the melting point of large sodium clusters [21]. The experiments by J. Parks *et al.* on trapped, size-selected clusters [169] suggests that electron diffraction could be used with great accuracy to measure transition temperatures, provided that an appropriate control of the cluster temperature is achieved. Electric properties, including susceptibilities and electric dipoles, are now being measured for various clusters ranging from metallic [170], ionic [69] to covalent [171] species. Their use as a possible probe of melting was demonstrated in sodium clusters by several authors [172], and we have shown here how the electric dipole itself could be employed for thermodynamical purposes. More generally, many techniques used to determine cluster structure could potentially be extended to the detection of phase changes.

The liquid-gas transition is farther from a complete understanding. This is mainly due to the difficulty in controlling the initial state of the cluster before fragmentation, as the cluster is most often heated by collisions. Some theoretical issues concerned with the observation time, the available volume, and the interpretation in terms of pressure [29] need more work to find useful indicators of boiling, and relevant definitions of the cluster boiling point. A related goal is to characterize the order of this finite-size transition, which is sometimes considered as a critical phenomenon, while it is well known to be first-order in the bulk. Noteworthy, no conclusive evidence of any second-order transition in clusters has been reported up to now.

Finally, many other kinds of temperature-induced transitions in clusters were not covered in this review. Transitions involving magnetic and nonmagnetic phases are a widely studied field, both experimentally and theoretically, in single nanoparticles as well as in nano-assemblies. The superfluid transition in helium clusters was studied by Jortner [173]. Phase changes in nanoparticles deposited on a substrate, inserted into a solid or a liquid medium are very important for practical applications, where it is highly desirable to predict the thermal stability of condensed phases at the nanoscale. Contrary to the experimental situation where the properties of the nanoparticles can be more easily tailored, atomistic modelling of clusters in their environment remains a difficult task when combined with statistical concerns. However, constant progresses in computational power and algorithms should lead to significant advances in the near future.

**Acknowledgments**

I would like to thank the people who have shared their ideas through many enjoying discussions in the recent years, in Toulouse (P. Labastie, F. Spiegelman, J. Vigué), Cambridge (J. P. K. Doye, D. J. Wales), Orsay (A. Boutin, A. H. Fuchs, P. Parneix, G. Torchet, M.-F. de Feraudy) and elsewhere (E. Yurtsever, D. L. Freeman, F. G. Amar, Ph. Dugourd, M. Broyer).

# References

[1] *Clusters of atoms and molecules, Vol. I*, edited by H. Haberland, Springer, Berlin, 1994.

[2] *Large clusters of atoms and molecules*, edited by T. P. Martin, Kluwer, Dordrecht, 1996.

[3] J. Jortner, *Z. Phys. D: At. Mol. Clusters* **24**, 247 (1992).

[4] T. L. Hill, *Thermodynamics of small systems*, Dover, New York, 1963.

[5] J. K. Lee, J. A. Barker, and F. F. Abraham, *J. Chem. Phys.* **58**, 3166 (1973).

[6] D. J. McGinty, *J. Chem. Phys.* **658**, 4733 (1973).

[7] C. L. Briant and J. J. Burton, *J. Chem. Phys.* **63** 2045 (1975).

[8] J. B. Kaelberer and R. D. J. Etters, *J. Chem. Phys.* **66**, 3233 (1977).

[9] V. V. Nauchitel and A. J. Pertsin, *Molec. Phys.* **40**, 1341 (1980).

[10] N. Quirke and P. Sheng, *Chem. Phys. Lett.* **110**, 63 (1984).

[11] R. S. Berry, J. Jellinek, and G. Natanson, *Chem. Phys. Lett.* **107**, 227 (1984).

[12] G. Natanson, F. Amar, and R. S. Berry, *J. Chem. Phys.* **78**, 399 (1983).

[13] T. L. Beck, J. Jellinek, and R. S. Berry, *J. Chem. Phys.* **87**, 545 (1987).

[14] T. L. Beck and R. S. Berry, *J. Chem. Phys.* **88**, 3910 (1988).

[15] T. L. Beck, D. M. Leitner, and R. S. Berry, *J. Chem. Phys.* **89**, 1681 (1988).

[16] R. S. Berry, *J. Phys. Chem.* **98**, 6910 (1994).

[17] P. Labastie and R. L. Whetten, *Phys. Rev. Lett.* **65**, 1567 (1990).

[18] M. Y. Hahn and R. L. Whetten, *Phys. Rev. Lett.* **61**, 1190 (1988).

[19] U. Even, N. Ben-Horin, and J. Jortner, *Phys. Rev. Lett.* **62**, 140 (1989).

[20] M. Schmidt, C. Ellert, W. Kronmüller, and H. Haberland, *Phys. Rev. B* **59**, 10970 (1999); M. Schmidt and H. Haberland, *Eur. Phys. J. D* **6**, 109 (1999).

[21] T. P. Martin, U. Näher, H. Schaber, and U. Zimmermann, *J. Chem. Phys.* **100**, 2322 (1994).

[22] M. Takagi, *J. Phys. Soc. Japan* **9**, 359 (1954).

[23] P. Buffat and J.-P. Borel, *Phys. Rev. A* **13**, 2287 (1976).

[24] Ph. Dugourd, R. R. Hudgins, and M. F. Jarrold, *Chem. Phys. Lett.* **267**, 186 (1997); R. R. Hudgins, Ph. Dugourd, J. M. Tenenbaum, and M. F. Jarrold, *Phys. Rev. Lett.* **78**, 4213 (1997).

[25] A. A. Shvartsburg and M. F. Jarrold, *Phys. Rev. Lett.* **85**, 2530 (2000).

[26] M. Schmidt, R. Kusche, W. Kronmüller, B. von Issendorff, and H. Haberland, *Phys. Rev. Lett.* **79**, 99 (1997).

[27] M. Schmidt, R. Kusche, B. von Issendorff, and H. Haberland, *Nature* (London) **393**, 238 (1998).

[28] M. Schmidt, R. Kusche, T. Hippler, J. Donges, W. Kronmüller, B. von Issendorff, and H. Haberland, *Phys. Rev. Lett.* **86**, 1191 (2001).

[29] M. Schmidt, T. Hippler, J. Donges, W. Kronmüller, B. von Issendorff, H. Haberland, and P. Labastie, *Phys. Rev. Lett.* **87**, 203402 (2001).

[30] J. Pochodzalla, T. Möhlenkamp, T. Rubehn *et al.*, *Phys. Rev. Lett.* **75**, 1040 (1995).

[31] C. Bréchignac, Ph. Cahuzac, B. Concina, and J. Leygnier, *Phys. Rev. Lett.* **89**, 203401 (2002).

[32] F. Gobet, B. Farizon, M. Farizon *et al.*, *Phys. Rev. Lett.* **89**, 183403 (2002).

[33] M. Chabot and K. Wohrer, *Phys. Rev. Lett.* **93**, 39301 (2004); F. Gobet , B. Farizon, M. Farizon *et al.*, *ibid.* **93**, 39302 (2004).

[34] C. E. Klots, *J. Chem. Phys.* **83**, 5854 (1985).

[35] B. Hartke, *Angew. Chem.* **41**, 1468 (2002).

[36] M. Bixon and J. Jortner, *J. Chem. Phys.* **91**, 1631 (1989).

[37] J. P. K. Doye, M. A. Miller, and D. J. Wales, *J. Chem. Phys.* **110**, 6896 (1999).

[38] D. J. Wales and J. P. K. Doye, *J. Phys. Chem. A* **101**, 5111 (1997).

[39] D. J. Wales, *Molec. Phys.* **78**, 151 (1993).

[40] J. P. K. Doye and D. J. Wales, *J. Chem. Phys.* **102**, 9659 (1995).

[41] F. Calvo, J. P. K. Doye, and D. J. Wales, *J. Chem. Phys.* **115**, 9627 (2001).

[42] J. P. K. Doye and D. J. Wales, *Phys. Rev. Lett.* **80**, 1357 (1998).

[43] R. H. Swendsen and J.-S. Wang, *Phys. Rev. Lett.* **57**, 2607 (1986).

[44] G. J. Geyer, in *Computing Science and Statistics: Proceedings of the 23rd Symposium on the Interface*, ed. by E. K. Keramidas (Interface Foundation, Fairfax Station, 1991), p. 156.

[45] J. P. Neirotti, F. Calvo, D. L. Freeman, and J. D. Doll, *J. Chem. Phys.* **112**, 10340 (2000).

[46] F. Calvo, J. P. Neirotti, D. L. Freeman, and J. D. Doll, *J. Chem. Phys.* **112**, 10350 (2000).

[47] P. J. Steinhardt, D. R. Nelson, and M. Ronchetti, *Phys. Rev. B* **28**, 784 (1983).

[48] C. L. Cleveland, W. D. Luedtke, and U. Landman, *Phys. Rev. Lett.* **81**, 2036 (1998).

[49] F. Calvo, J. P. K. Doye, and D. J. Wales, *Phys. Rev. Lett.* **87**, 119301 (2001).

[50] L. A. Girifalco, *J. Phys. Chem.* **96**, 858 (1992).

[51] F. Calvo, J. P. K. Doye, and D. J. Wales, *J. Chem. Phys.* **114**, 7132 (2001).

[52] F. Calvo and J. P. K. Doye, *Phys. Rev. E* **63**, 010902 (2000).

[53] D. D. Frantz, *J. Chem. Phys.* **102**, 3747 (1995).

[54] D. D. Frantz, J. Chem. Phys. **115**, 6136 (2001).

[55] F. Calvo, F. Spiegelman, and M.-C. Heitz, *J. Chem. Phys.* **118**, 8739 (2003).

[56] F. Calvo, F. Spiegelman, and D. J. Wales, *J. Chem. Phys.* **118**, 8754 (2003).

[57] O. M. Becker and M. Karplus, *J. Chem. Phys.* **106**, 1495 (1997).

[58] D. J. Wales, J. P. K. Doye, M. A. Miller, P. N. Mortenson, and T. R. Walsh, *Adv. Chem. Phys.* **115**, 1 (2000).

[59] D. J. Wales, Energy Landscapes, Cambridge University, Cambridge, 2003.

[60] S. Krivov and M. Karplus, *J. Chem. Phys.* **117**, 10894 (2002).

[61] D. A. Evans and D. J. Wales, *J. Chem. Phys.* **118**, 3891 (2003).

[62] C. J. Burnham, S. S. Xantheas, M. A. Miller, B. E. Applergate, and R. E. Miller, *J. Chem. Phys.* **117**, 1109 (2002).

[63] T. P. Martin, *Phys. Rep.* **95**, 167 (1983).

[64] J. Luo, U. Landman, and J. Jortner, in *Physics and Chemistry of Small Clusters*, NATO ASI series B, edited by P. Jena, B. Rao, and K. Khanna (Plenum, New York 1987), vol 158, p 155.

[65] A. Heidenreich, I. Oref, and J. Jortner, J. Phys. Chem. **91**, 4950 (1987); A. Heidenreich, I. Schek, D. Scharf, and J. Jortner, *Z. Phys. D: At. Mol. Clusters* **20**, 227 (1991).

[66] J. P. K. Doye and D. J. Wales, *Phys. Rev. B* **59**, 2292 (1999).

[67] F. Calvo and P. Labastie, *J. Phys. Chem. B* **102**, 2051 (1998).

[68] J. P. K. Doye and D. J. Wales, *J. Chem. Phys.* **111**, 11070 (1999).

[69] D. Rayane, I. Compagnon, R. Antoine, M. Broyer, P. Dugourd, P. Labastie, J.-M. L'Hermite, A. Le Padellec, G. Durand, F. Calvo, F. Spiegelman, and A. R. Allouche, *J. Chem. Phys.* **116**, 10730 (2002).

[70] V. Kumar and Y. Kawazoe, *Phys. Rev. B* **66**, 14413 (2002).

[71] P. Nava, M. Sierka, and R. Ahlrichs, *Phys. Chem. Chem. Phys.* **5**, 3372 (2003).

[72] C. Rey, L. J. Gallego, J. García-Rodeja, J. A. Alonso, M. P. Iñiguez, *Phys. Rev. B* **48**, 8253 (1993).

[73] D. Tománek, Z. Sun, and S. G. Louie, *Phys. Rev. B* **43**, 4699 (1991).

[74] W. Zhong, Y. S. Li, and D. Tománek, *Phys. Rev. B* **44**, 13053 (1991).

[75] W. Zhong, Y. Cai, and D. Tománek, *Phys. Rev. B* **46**, 8099 (1992).

[76] H. Grönbeck, D. Tománek, S. G. Kim, and A. Rosén, *Z. Phys. D: At. Mol. Clusters* **40**, 469 (1997); *Chem. Phys. Lett.* **264**, 39 (1997).

[77] F. Calvo and A. Carré, to be published.

[78] C. Molteni, R. Martoňák, and M. Parrinello, *J. Chem. Phys.* **114**, 5358 (2001).

[79] R. Martoňák, L. Colombo, C. Molteni, and M. Parrinello, *J. Chem. Phys.* **117**, 11329 (2002).

[80] C. B. Barber, D. P. Dobkin, and H. Huhdanpaa, *ACM T. Math. Software* **22**, 469 (1996).

[81] J. P. K. Doye, D. J. Wales, F. H. M. Zetterling, and M. Dzugutov, *J. Chem. Phys.* **118**, 2792 (2003).

[82] F. Calvo and J. P. K. Doye, *Phys. Rev. B* **69**, 125414 (2004).

[83] J. P. K. Doye and F. Calvo, *Phys. Rev. Lett.* **86**, 3570 (2001).

[84] J. P. K. Doye and F. Calvo, *J. Chem. Phys.* **116**, 8307 (2002).

[85] F. Baletto, C. Mottet, and R. Ferrando, *Surf. Sci.* **446**, 31 (2000).

[86] J. P. K. Doye, D. J. Wales, and R. S. Berry, *J. Chem. Phys.* **103**, 4234 (1995).

[87] H.-P. Cheng, X. Li, R. L. Whetten, and R. S. Berry, *Phys. Rev. A* **46**, 791 (1992).

[88] R. Poteau, F. Spiegelmann, and P. Labastie, *Z. Phys. D: At. Mol. Clusters* **30**, 57 (1994).

[89] S. G. Kim and D. Tománek, *Phys. Rev. Lett.* **72**, 2418 (1994).

[90] A. Aguado, J. M. López, J. A. Alonso, and M. J. Stott, *J. Chem. Phys.* **111**, 6026 (1999); *J. Phys. Chem. B* **105**, 2386 (2001).

[91] K. Joshi, D. G. Kanhere, and S. A. Blundell, *Phys. Rev. B* **66**, 155329 (2002); **67**, 235413 (2003).

[92] A. Rytkönen, H. Häkkinen, and M. Manninen, *Eur. Phys. J. D* **8**, 93 (2000); A. Rytkönen and M. Manninen, *ibid.* **23**, 351 (2003).

[93] J. P. Rose and R. S. Berry, *J. Chem. Phys.* **96**, 517 (1991).

[94] J. P. Rose and R. S. Berry, *J. Chem. Phys.* **98**, 3246 (1992).

[95] J. P. Rose and R. S. Berry, *J. Chem. Phys.* **98**, 3262 (1992).

[96] B. Vekhter and R. S. Berry, *J. Chem. Phys.* **106**, 6456 (1997).

[97] E. M. Pearson, T. Halicioglu, and W. A. Tiller, *Phys. Rev. A* **32**, 3030 (1985).

[98] D. J. Wales and R. S. Berry, *Phys. Rev. Lett.* **73**, 2875 (1994).

[99] R. M. Lynden-Bell and D. J. Wales, *J. Chem. Phys.* **101**, 1460 (1994).

[100] W. Thirring, H. Narnhofer, and H. A. Posch, *Phys. Rev. Lett.* **91**, 130601 (2003).

[101] D. J. Wales and J. P. K. Doye, *J. Chem. Phys.* **103**, 3061 (1995).

[102] G. A. Breaux, R. C. Benirschke, T. Sugai, B. S. Kinnear, and M. F. Jarrold, *Phys. Rev. Lett.* **91**, 215508 (2003).

[103] K. Joshi, D. G. Kanhere, and S. A. Blundell, *Phys. Rev. B* **67**, 235413 (2003); C. Majumder, V. Kumar, H. Mizuseki, and Y. Kawazoe, *Phys. Rev. B* **64**, 233405 (2001).

[104] S. Chacko, K. Joshi, D. G. Kanhere, and S. A. Blundell, *Phys. Rev. Lett.* **92**, 135506 (2004).

[105] M. Schmidt, J. Donger, T. Hippler, and H. Haberland, *Phys. Rev. Lett.* **90**, 103401 (2003).

[106] H. Haberland, T. Hippler, J. Donges, O. Kostko, M. Schmidt, and B. von Issendorff, *Phys. Rev. Lett.* **94**, 035701 (2005).

[107] M. Moseler, H. Häkkinen, and U. Landman, *Phys. Rev. Lett.* **87**, 053401 (2001).

[108] F. Calvo and F. Spiegelman, *Phys. Rev. Lett.* **82**, 2270 (1999).

[109] F. Calvo and F. Spiegelman, *J. Chem. Phys.* **112**, 2088 (2000).

[110] F. Calvo and F. Spiegelman, *Phys. Rev. Lett.* **89**, 266401 (2002).

[111] F. Calvo and F. Spiegelman, *J. Chem. Phys.* **120**, 9684 (2004).

[112] Z. Li and H. A. Scheraga, *Proc. Nat. Acad. Sci. U.S.A.* **84**, 6611 (1987).

[113] J. A. Reyes-Nava, I. L. Garzón, M. R. Beltrán, and K. Michaelian, *Rev. Mex. Fis.* **48**, 450 (2002); J. A. Reyes-Nava, I. L. Garzón, and K. Michaelian, *Phys. Rev. B* **67**, 165401 (2003).

[114] Y. Li, E. Blaisten-Barojas, and D. A. Papaconstantopoulos, *Chem. Phys. Lett.* **268**, 331 (1997).

[115] R. Poteau and F. Spiegelmann, *Phys. Rev. B* **45**, 1878 (1991).

[116] R. Poteau and F. Spiegelmann, *J. Chem. Phys.* **98**, 6540 (1993).

[117] F. Calvo, S. Tran, S. A. Blundell, C. Guet, and F. Spiegelmann, *Phys. Rev. B.* **62**, 10394 (2000).

[118] H.-P. Cheng and R. S. Berry, *Phys. Rev. A* **45**, 7969 (1992).

[119] F. Calvo and P. Labastie, *Chem. Phys. Lett.* **248**, 233 (1996).

[120] J. P. K. Doye and D. J. Wales, *Z. Phys. D* **40**, 466 (1997).

[121] A. Bulgac and C. Lewenkopf, *Europhys. Lett.* **31**, 519 (1995).

[122] M. R. Ghayal and E. Curotto, *J. Chem. Phys.* **111**, 5522 (1999).

[123] A. M. Aguado and E. Curotto, *Chem. Phys. Lett.* **330**, 440 (2000).

[124] J. H. Skone and E. Curotto, *J. Chem. Phys.* **117**, 7137 (2002).

[125] M. A. Gaveau, M. Briant, P. R. Fournier, J. M. Mestdagh, J. P. Visticot, F. Calvo, S. Baudrand, and F. Spiegelman, *Euro. Phys. J. D* **21**, 153 (2002).

[126] J. M. Mestdagh, M. A. Gaveau, C. Gée, O. Sublimontier, and J. P. Visticot, *Int. Rev. Phys. Chem.* **16**, 215 (1997).

[127] F. Calvo, F. Spiegelman, and J.-M. Mestdagh, *J. Chem. Phys.* **118**, 8763 (2003).

[128] R. P. Feynman and A. R. Hibbs, *Quantum mechanics and path integrals* (McGraw-Hill, New York, 1965).

[129] H. Wadi and E. Pollak, *J. Chem. Phys.* **110**, 11890 (1999).

[130] S. Sastry and C. Angell, *Nature Materials*, **2**, 739 (2003).

[131] A. Boutin, B. Rousseau, and A. H. Fuchs, Chem. Phys. Lett. **218**, 122 (1994); A. Boutinn, J.-B. Maillet, and A. H. Fuchs, *J. Chem. Phys.* **99**, 9944 (1993).

[132] D. J. Wales and I. Ohmine, *J. Chem. Phys.* **98**, 7245 (1993); **98**, 7257 (1993).

[133] C. J. Tsai and K. D. Jordan, *J. Chem. Phys.* **99**, 6957 (1993).

[134] G. Del Mistro and A. J. Stace, *Chem. Phys. Lett.* **171**, 281 (1990).

[135] F. J. Dulles and L. S. Bartell, *J. Phys. Chem.* **99**, 17100 (1995); L. S. Bartell and F. J. Dulles, *J. Phys. Chem.* **99**, 17107 (1995).

[136] R. D. Etters, K. Flurchick, R. P. Pan, and V. Chandrasekharan, *J. Chem. Phys.* **75**, 929 (1981).

[137] J.-B. Maillet, A. Boutin, and A. H. Fuchs, *Phys. Rev. Lett.* **76**, 4336 (1996); J.-B. Maillet, A. Boutin, S. Buttefey, F. Calvo, and A. H. Fuchs, *J. Chem. Phys.* **109**, 329 (1998).

[138] P. Parneix, *Euro. Phys. J. D* **23**, 375 (2003).

[139] M. L. Klein, D. Levesque, and J.-J. Weis, *Phys. Rev. B* **21**, 5785 (1980).

[140] R. D. Etters, K. Kobashi, and J. Belak, *Phys. Rev. B* **32**, 4097 (1985); R. LeSar and R. D. Etters, *ibid.* **37**, 5364 (1987).

[141] F. Calvo, *Phys. Rev. E* **60**, 2771 (1999).

[142] M. Moseler and J. Nordiek, *Phys. Rev. B* **60**, 11734 (1999).

[143] F. Calvo, *J. Phys. Chem. B* **105**, 2183 (2001).

[144] W. Forst, Unimolecular Reactions, Cambridge University Press, Cambridge, 2003.

[145] R. G. Gilbert and S. C. Smith, *Theory of Unimolecular and Recombination Reactions*, Blackwell Scientific, Oxford, 1990.

[146] M. F. Jarrold, Introduction to statistical reaction theories, in *Clusters of Atoms and Molecules I*, edited by H. Haberland, Springer, Berlin, 1991.

[147] S. Weerasinghe, F. G. Amar, *Z. Phys. D: At., Mol. Clusters* **20**, 167 (1991); *J. Chem. Phys.* **98**, 4967 (1993).

[148] M. Belkacem, V. Latora, and A. Bonasera, *Phys. Rev. C* **52**, 271 (1995).

[149] X. Campi and H. Krivine, *Nucl. Phys. A* **620**, 46 (1997).

[150] Ph. Chomaz, V. Duflot, and F. Gulminelli *Phys. Rev. Lett.* **85**, 3587 (2000).

[151] M. E. Fisher, *Rep. Prog. Phys.* **30**, 615 (1967).

[152] D. H. E. Gross, *Rep. Prog. Phys.* **53**, 605 (1990).

[153] G. Martinet *et al.*, *Euro. Phys. J. D* **24** (2003) 149.

[154] A. Strachan and C. O. Dorso, *Phys. Rev. C* **59**, 285 (1999).

[155] O. K. Rice and H. C. Ramsperger, *J. Am. Chem. Soc.* **50**, 617 (1928); L. S. Kassel, *J. Phys. Chem.* **32**, 225 (1928).

[156] V. Weisskopf, *Phys. Rev.* **52**, 295 (1937).

[157] P. C. Engelking, *J. Chem. Phys.* **85**, 3103 (1986); **87**, 936 (1987).

[158] E. Nikitin, *Theor. Exp. Chem.* **1**, 83 (1965); **1**, 90 (1965).

[159] P. Pechukas and J. C. Light, *J. Chem. Phys.* **42**, 3281 (1965).

[160] C. E. Klots, *J. Phys. Chem.* **75**, 1526 (1971).

[161] W. J. Chesnavich and M. T. Bowers, *J. Chem. Phys.* **66**, 2306 (1977).

[162] F. Calvo and P. Parneix, *J. Chem. Phys.* **119**, 256 (2003).

[163] P. Parneix and F. Calvo, *J. Chem. Phys.* **119**, 9469 (2003).

[164] F. Calvo and P. Parneix, *J. Chem. Phys.* **120**, 2780 (2004).

[165] P. Parneix, Ph. Bréchignac, and F. Calvo, *Chem. Phys. Lett.* **381**, 471 (2003).

[166] I. Napari and H. Vehkamäki, *J. Chem. Phys.* **121**, 819 (2004).

[167] C. Caccamo, D. Costa, and A. Fucile, *J. Chem. Phys.* **106** 255 (1997).

[168] B. Schmidt and R. B. Gerber, *Phys. Rev. Lett.* **72**, 2490 (1994).

[169] M. Maier-Borst, D. B. Cameron, M. Rokni, and J. H. Parks, *Phys. Rev. A* **59**, R3162 (1999); S. Kruckeberg, D. Schooss, M. Maier-Borst and J. H. Parks, *Phys. Rev. Lett.* **85**, 4494 (2000).

[170] R. Antoine, D. Rayane, A. R. Allouche, M. Aubert-Frécon, E. Benichou, F. W. Dalby, Ph. Dugourd, M. Broyer, and C. Guet, *J. Chem. Phys.* **110**, 5568 (1999).

[171] R. Antoine, Ph. Dugourd, D. Rayane, E. Benichou, M. Broyer, F. Chandezon, and C. Guet, *J. Chem. Phys.* **110**, 9771 (1999).

[172] S. Kümmel, J. Akola, and M. Manninen, *Phys. Rev. Lett.* **84**, 3827 (2000); L. Kronik, I. Vasiliev, M. Jain, and J. R. Chelikowsky, *J. Chem. Phys.* **115**, 4322 (2001).

[173] J. Jortner, *J. Chem. Phys.* **119**, 11335 (2003).

In: Progress in Chemical Physics Research
Editor: A. N. Linke, pp. 65-80
ISBN 1-59454-451-4


*Chapter 3*

# CHIRALITY SELECTION MODELS IN A CLOSED SYSTEM

***Yukio Saito and Hiroyuki Hyuga***
Department of Physics, Keio University,
Yokohama 223-8522, Japan

**Abstract**

Recently the first examples of chemical systems which show the amplification of enantiomeric excess (ee) were found. Inspired by these experiments, we have previously proposed a chemical reaction model which leads to chirality selection in a closed system. Here we extend the model and analyze general ones in detail by means of flow diagrams in the phase space of enantiomer concentrations.

With only a quadratic autocatalytic process in the production of chiral molecules from achiral substrate, ee is found to be amplified, but the flow is attracted to a fixed line and the final degree of ee depends strongly on the initial condition. When the produced chiral molecules decompose back to achiral substrate, they are recycled in the catalytic production, because the system is closed. Various recycling processes are found to alter the final state of the system completely: With a cross inhibition and a linear autocatalysis a chiral symmetry of the system is broken and a complete homochirality is achieved. With a back reaction only, the quadratic autocatalysis is found necessary to establish chirality selection.

## 1 Introduction

When a carbon atom in an organic molecule extends its four chemical bonds towards the direction of tetrahedral corners, the molecule takes a three-dimensional structure. If those bonds are connected to four different chemical species, the molecule has two possible stereostructures which cannot be overlapped each other by simple rotations; Only a mirror image of one form overlaps with another, like the left and right hands. [1, 2]. The molecule is called chiral, and the two forms are enantiomers with each other. In general the two enantiomers are classified in $R$ and $S$ forms, but, for sugars and amino acids, D and L representations are common in use. From energetic point of view, the two enantiomers are indistinguishable and they should exist with equal probability. In fact, in the usual chemical production from achiral substrates both types of enantiomers are produced with equal probabilities, and the product is called a racemate or a racemic mixture.

In contrast, life on earth utilizes only one type of enantiomers: levorotatory(L)-amino acids and dextrorotatory(D)-sugars. This symmetry breaking in chirality is called the homochirality, and the origin of this unique chirality selection has long been intriguing many scientists [1]. Various mechanisms have been proposed to induce an asymmetry in the primordial molecular environment [3, 4, 5, 6, 7, 2, 8], but the induced chiral asymmetry in all the proposed models is expected to be very minute. Therefore, it is indispensable to amplify the enantiomeric excess (ee).

Frank is the first to show that an autocatalytic chemical reaction with an antagonistic process can lead to an ee amplification in open systems [9]. Many theoretical models have since been proposed, but they are often criticized as lacking any experimental support [1]. Recently, a first example of a system which shows the spontaneous amplification of ee has been found by Soai's group [10, 11, 12, 13]. This enhancement takes place in a closed chemical reaction system, and is explained by the second-order autocatalytic reaction[13]. Inspired by these experiments [10, 11, 12, 13], we constructed some simple models of autocatalytic chemical reactions in a closed and conserved system [14, 15], and searched conditions not only for the ee amplification but also for the unique chirality selection. Here we extend these models and study them in detail.

In §2, we summarize the fundamental aspects of autocatalytic reaction models where two enantiomers compete against each other in a closed system for the limited resource [14]. The ee amplification is found possible if the production of chiral molecules contains the quadratic autocatalytic process. However, with only the autocatalyses, the final concentrations of enantiomers are not unique but depend strongly on the initial condition. For the selection of a unique final state, the recycling proceses are reqiured as discussed in §3 and 4: In §3, by including the mutual antagonism or the cross inhibition, a linear autocatalysis is shown to be sufficient to establish a complete homochirality. In §4, with the simplest back reaction, the quadratic autocatalysis is found necessary for the chirality selection. Summary and discussions are given in the last section §5.

## 2 Autocatalytic Reactions in a Closed System

Frank [9] has considered chiral symmetry breaking in an open chemical system. However, the experiment is performed in a closed system [11], and we think it necessary to find a chemical reaction model which shows chirality selection in a closed system. In order to focus on the effects of the autocatalysis on ee, we start our study from those reaction models with only autocatalytic productions of chiral molecules from achiral substrates.

Suppose two achiral molecules A and B react to produce a chiral molecule C. For the purpose of simplicity we denote an enantiomer ($R$)-C molecule as R and the other ($S$)-C molecule as S, hereafter. The reactions A+B→ R and A+B→ S proceed in a closed system, and thus they stop when the materials A and/or B are exhausted. Here we assume that there is an ample amount of material B, and its concentration can be regarded as constant all through the reaction. Then, the process depends only on the concentrations of A, R and S molecules, which we denote $a$, $r$ and $s$, respectively. The conservation of the total number of participating molecules gives the relation that the total concentration stays constant determined by the initial values as

$$a + r + s = a_0 + r_0 + s_0 = c_0 = \text{const.} \tag{1}$$

Then the developement of the reaction can be depicted as a flow in the $r$-$s$ phase space. The conservation law (1) limits the phase space in the triangular region

$$0 \leq r, s \leq c_0, \quad 0 \leq r + s \leq c_0. \tag{2}$$

As a first step we consider the simplest autocatalytic reaction such that the production from the substrate A to R or S is described by the following rate equations:

$$\begin{aligned} \frac{dr}{dt} &= f(r)a = f(r)(c_0 - r - s) \\ \frac{ds}{dt} &= f(s)a = f(s)(c_0 - r - s) \end{aligned} \tag{3}$$

with

$$f(x) = k_0 + k_1 x + k_2 x^2. \tag{4}$$

The coefficient $k_0$ represents the rate of a nonautocatalitic reaction, $k_1$ of a linearly autocatalytic and $k_2$ of a quadratically autocatalytic reactions. Since all the rates $k_{0,1,2}$ are non-negative, $f(x) \geq 0$ for $x = r$ and $s$. In the Soai reaction, the quadratic autocatalysis is shown to explain the time evolution of the reaction yield and the ee [13].

## 2.1 Fixed Points and Flow Diagram

We now study the flow diagram in $r - s$ space in a few typical situations. First we search fixed points of Eq.(3) in the region (2). It is easy to find that $a = 0$, where all the substrate A is consummed up, gives a line of fixed points at $r + s = c_0$. Only for the case without nonautocatalysis, $k_0 = 0$, another fixed point is possible at the origin $r = s = 0$, but it is easily shown to be unstable.

If the initial state does not lie on neither fixed points nor fixed lines, the enantiomer concentrations evolve as functions of time. The evolution is depicted in a flow diagram in $r$-$s$ phase space. In the present case, the flow is attracted to the fixed line $r + s = c_0$. The flow trajectory is given analytically by the equation

$$\int \frac{dr}{f(r)} - \int \frac{ds}{f(s)} = \text{constant}. \tag{5}$$

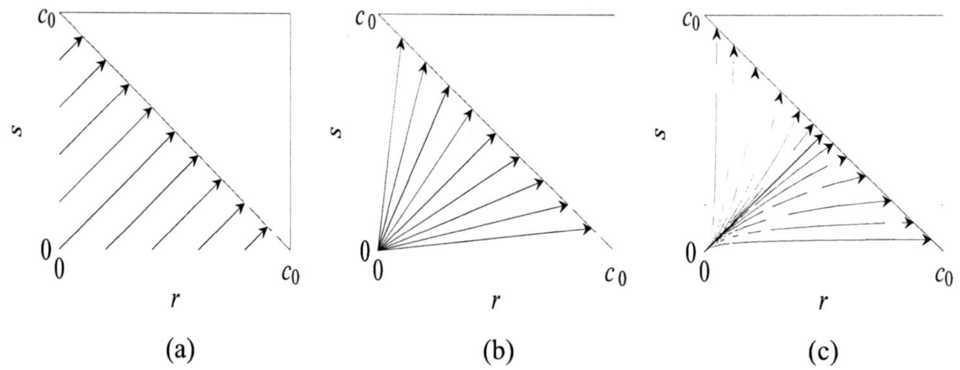


Figure 1: Flow diagrams for (a) nonautocatalytic ($k_0 > 0,\ k_1 = k_2 = 0$), (b) linearly autocatalytic ($k_1 > 0,\ k_0 = k_2 = 0$), and (c) quadratically autocatalytic ($k_2 > 0,\ k_0 = k_1 = 0$) reactions.

It can be integrated analytically for general $f(x)$, but we discuss the flow lines only in three typical cases in the following.

• **Nonautocatalytic case:**

Without autocatalysis, $k_0 > 0$ but $k_1 = k_2 = 0$, integration of eq.(5) is straightforward and yields flow lines with a unit slope, $r - s$ =constant, as shown in Fig.1(a).

• **Linearly autocatalitic case:**

With only a linear autocatalysis, $k_1 > 0,\ k_0 = k_2 = 0$, the integration of eq.(5) yields the trajectory as $r/s$ =constant. The flow is represented by lines projecting out radially from the origin, as shown in Fig.1(b).

• **Quadratically autocatalytic case:**

For a quadratically autocatalytic case, $k_2 > 0,\ k_0 = k_1 = 0$, flow lines are a collection of hyperbolae projecting out from the origin, $1/r - 1/s$ =constant, as shown in Fig.1(c).

## 2.2 Enantiomeric Excess

In order to quantify the ee or the degree of chiral symmetry breaking, we use an order parameter $\phi$ defined by

$$\phi = \frac{r - s}{r + s}, \tag{6}$$

which is zero for symmetric or racemic state ($r = s$), and is nonzero for a symmetry-broken or chiral state.

• **Nonautocatalytic case:**

For the non-autocatalytic case in Fig.1(a), the numerator $r - s$ of $\phi$ remains constant during the time evolution whereas the denominator $r + s$ increases. Thus the ee decreases monotonously from the initial value $|\phi_0| = |(r_0 - s_0)/(r_0 + s_0)|$ to the asymptotic one

$|\phi_\infty| = |(r_0 - s_0)/c_0|$, since $r + s = c_0 - a$ approaches $c_0$ asymptotically. If the initial concentrations $r_0$ and $s_0$ are small, ee or $|\phi|$ becomes very small. This result is quite natural from the microscopic point of view, since the nonautocatalytic reaction produces both enantiomers randomly and leads to the racemization.

• **Linearly autocatalytic case:**

For the linearly autocatalytic case in Fig.1(b), the ratio $r/s$ remains constant and so does the ee or $\phi$.

• **Quadratically autocatalytic case:**

Only with a nonlinear autocatalysis in Fig.1(c), the ee increases up to some saturation value $|\phi_\infty|$, as shown in Fig.2. The asymptotic value $\phi_\infty$ depends on the initial concentrations $r_0$ and $s_0$, even when the initial $\phi_0$ and the total $c_0$ are fixed, as shown in Fig.2. The smaller the initial amount of chiral molecules $(r_0 + s_0)$ are, the more the ee is amplified.

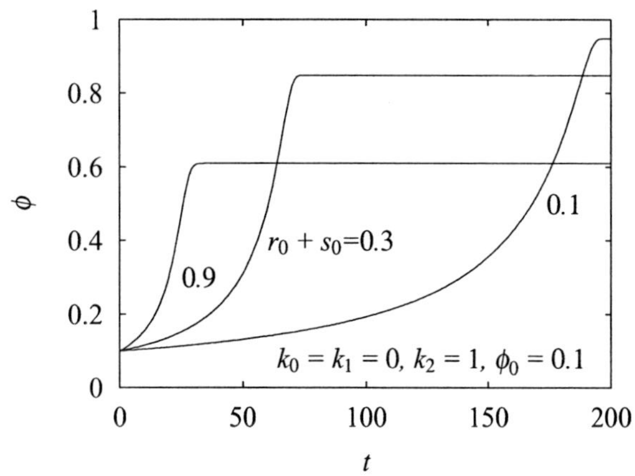


Figure 2: Time evolution of the enantiomeric excess $\phi$ with quadratic autocatalytic reaction $k_2 > 0$ with different amount of initial R and S molecules, $r_0$ and $s_0$. The initial value $\phi_0 = 0.1$ and the total concentration $c_0 = 1$ are kept the same.

• **General case:**

In general, the ee amplification is explained by studying the evolution of $\phi$, which is derived from eq.(3) as

$$\frac{d\phi}{dt} = \frac{a}{c_0 - a}\phi[-2k_0 + \frac{k_2}{2}(c_0 - a)^2(1 - \phi^2)]. \tag{7}$$

The evolution equation turns out to be independent of the linear autocatalytic rate $k_1$: It is contained in the time evolution of $a$. Without quadratic autocatalysis ($k_2 = 0$) the ee decreases as we have already observed, since the concentration of A material, $a$, and the sum of R and S products, $c_0 - a = r + s$, should always be positive. The situation does not alter by including only a linear autocatalysis with $k_1 > 0$, because it does not directly affect the evolution of $\phi$. Only when the quadratic auatocatalysis dominates at $k_2 > 4k_0/c_0^2$, the ee can increase. Therefore, in this simplest model only the quadratic autocatalysis can enhance the chiral asymmetry. This may correspond to the situation observed in the Soai reaction [11, 12, 13]. It has to be noted that the asymptotic value $\phi_\infty$ cannot be determined from eq.(7), since $\lim_{t\to\infty} a(t) = 0$ in the present model. Even though ee amplification or the amplification of the chiral asymmetry is brought about by the nonlinear autocatalysis, the final state of the system is on the line of fixed points, and thus the final values of $\phi$ can be anything depending on the initial condition. These points are marginally stable because they have no quaranteed stability along the direction of the fixed line. Thus, they may be susceptible to additional perturbations. We now consider the effect of two types of destructive reactions of chiral molecules separately in the following two sections.

## 3 With Cross Inhibitions

Frank has introduced the mutual antagonism between the enantiomers, and succeeded in explaining the chiral symmetry breaking in an open system [9]. Here we modify his model adapted to the closed system. The rate equations now incorporate terms of cross inhibition as

$$\frac{dr}{dt} = f(r)a - \mu rs$$
$$\frac{ds}{dt} = f(s)a - \mu rs, \tag{8}$$

with the restriction that the total concentrations should be constant; $a + r + s = c_0$. Therefore, when enantiomers R and S react to decompose, two substrate molecules A are produced, which are further recycled in the production of chiral molecules.

### 3.1 Fixed Points

Fixed points have to satisfy the relation

$$f(r)a = f(s)a = \mu rs. \tag{9}$$

The simplest solution is $a = 0$ and $rs = 0$, which corresponds to the corners of the triangular region, $(r, s) = (c_0, 0)$ and $(0, c_0)$. These are easily shown to be stable with complete homochirality, $\phi = \pm 1$. Other fixed points are also possible with a nonzero $a$ and $f(r) = f(s) = \mu rs/a$. We shall study them in various cases below.

**• Nonautocatalytic case:**

Only with a nonautocatalytic reaction ($k_0 > 0,\ k_1 = k_2 = 0$), $f(r) = f(s) = k_0$ and there is a fixed curve of hyperbola

$$(r + \frac{k_0}{\mu})(s + \frac{k_0}{\mu}) = \frac{k_0}{\mu}(c_0 + \frac{k_0}{\mu}) \tag{10}$$

passing through the two terminal points $(r,\ s) = (c_0,\ 0)$ and $(0,\ c_0)$, as shown in Fig.3(a). The fixed curve, in fact, is an accumulation of infinitely many marginal fixed points. The flow is the same which is obtained in the previous section without cross inhibition: $r - s$ =constant. The cross inhibition only shifts the fixed line from the border $r + s = c_0$ in Fig.1(a) to the hyperbola (10) in Fig.3(a).

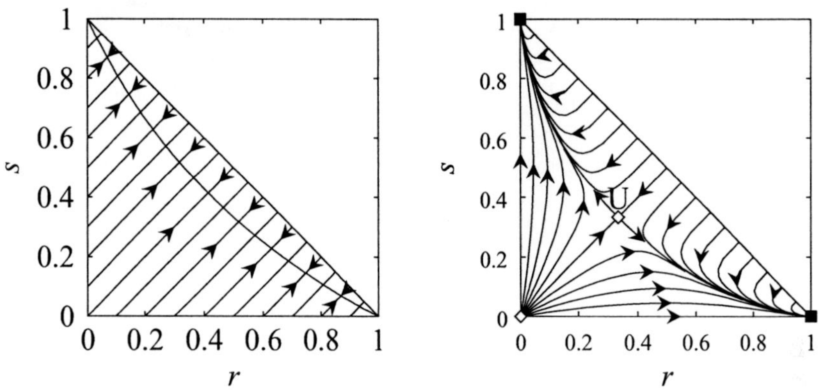


(a) (b)

Figure 3: Flow diagrams with a cross inhibition $\mu c_0 = 1$ at $c_0 = 1$ and (a) $k_0 > 0,\ k_1 = k_2 = 0$ and (b) $k_1 > 0,\ k_0 = k_2 = 0$. Fixed points marked by filled squares are stable, whereas those by open diamonds are unstable.

- **Linearly autocatalytic case:**

With a finite $k_1$ as well as $k_0$, a fixed line is replaced by a symmetric fixed point U at

$$r = s = \frac{-2k_0 + k_1 c_0 + \sqrt{(2k_0 + k_1 c_0)^2 + 4k_0 c_0 \mu}}{2(\mu + 2k_1)}. \tag{11}$$

For $k_0 = 0$, there appears an additional fixed point at the origin $r = s = 0$. All these fixed points are easily shown to be unstable, as is illustrated in Fig.3(b). The only stable fixed points are those at the corners, $(c_0, 0)$ and $(0, c_0)$, with the complete homochirality. Thus, with only a linear autocatalysis and with a cross inhibition, complete homochirality is possible in a closed system. Additional quadratic autocatalysis does not alter the overall structure of the flow.

## 3.2 Enantiomeric Excess

Up to the linear autocatalysis ($k_2 = 0$) with a cross inhibition ($\mu \neq 0$), the ee follows the evolution equation

$$\frac{d\phi}{dt} = \frac{1}{c_0 - a}\phi[-2k_0 a + \frac{\mu}{2}(c_0 - a)^2(1 - \phi^2)]. \tag{12}$$

The rate $k_1$ of the linear autocatalytic process is not involved in the symmetry breaking effect of $\phi$ irrespective of cross inhibition. One notices the similarity between the evolution eq.(12) and eq.(7) derived in the previous section. The effect of the quadratic autocatalytic reaction, $k_2$, in eq.(7) is replaced by $\mu/a$ in eq.(12). In the present case, the symmetry recovering effect $k_0$ weakens as the material A is consumed ($a \to 0$), and the complete homochirality $\phi = \pm 1$ is achieved asymptotically. For example, the time evolution of $r/c_0$, $s/c_0$ and $\phi$ is depicted in Fig.4 at $k_0 = 0, k_1 c_0 = 1, \mu c_0 = 1$ and $r_0/c_0 = 0.0714, s_0/c_0 = 0.0700, \phi_0 = 0.01$. Initially $r$ and $s$ increase simultaneously to approach the symmetric fixed point U: ($r/c_0$, $s/c_0$)=(1/3,1/3) by keeping $\phi$ close to $\phi_0$, driven by the linear autocatalytic process $k_1$. However, since U is unstable, the trajectory is eventually expelled from U and $r$ increases at the cost of $s$. As a result $\phi$ increases to the full homochirality.

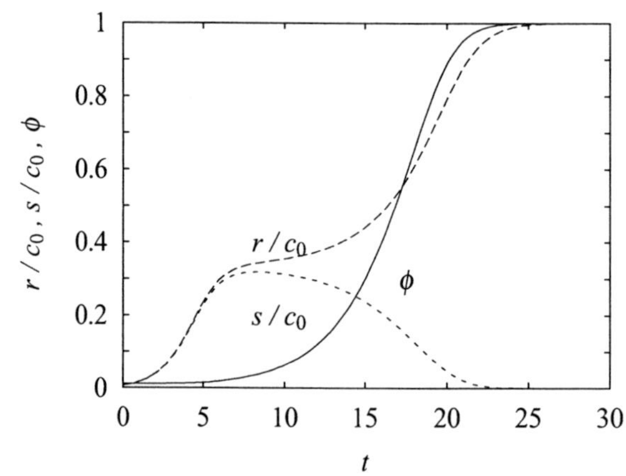


Figure 4: Time evolution of $r, s$, and $\phi$ at $k_0 = 0, k_1 c_0 = 1, \mu c_0 = 1$ and $r_0/c_0 = 0.0714, s_0/c_0 = 0.0700, \phi_0 = 0.01$.

# 4 With Back Reactions

We now consider another type of decomposition process, a simple back reaction from the product R or S to an A molecule. The back reaction is linear in the product concentrations $r$ and $s$, and governing rate equations are

$$\begin{aligned} \dot{r} &= f(r)a - \lambda r \\ \dot{s} &= f(s)a - \lambda s. \end{aligned} \tag{13}$$

Those substrate A molecules produced by the back reaction are recycled in the production of chiral molecules in a closed system.

In the present case with a linear recycling process, the evolution of $\phi$ is the same with eq.(7). Therefore, the quadratic autocatalysis with a finite $k_2$ is the necessity to break the chiral symmetry. Though $\phi$ follows the same evolution (7), the asmptotic behavior with a recycling is quite different from that without the back reaction in §2, because the fixed point structure is drastically altered by the back reaction.

## Fixed Points, Flow Diagram and Enantiomeric Excess

Fixed points of the flow satisfy conditions

$$f(r)a - \lambda r = f(s)a - \lambda s = 0. \tag{14}$$

It is trivial that $a = 0$ cannot be a fixed point any more unless $c_0 = 0$. As we have supposed, the fixed line $r + s = c_0$ in §2 is very sensible to the perturbation like a back reaction. We now study structures of fixed points in some simple cases.

**• Nonautocatalytic case:**

In the non-autocatalytic case with $k_0 > 0,\ k_1 = k_2 = 0$ with a finite $\lambda$, there is only a symmetric fixed point at

$$r = s = \frac{c_0/2}{1 + \lambda/2k_0} < \frac{c_0}{2}, \tag{15}$$

which is easily shown to be stable and is illustrated in Fig.5(a).

**• Linearly autocatalytic case:**

With only a linear autocatalysis $k_1 > 0,\ k_0 = k_2 = 0$, we have only a trivial fixed point at the origin $(r, s) = (0, 0)$, when the rate of back reaction is as large as $\lambda \geq k_1 c_0$: there will be no net production of chiral C molecules because of strong back reaction. On the contrary, when the rate of back reaction is small as $\lambda < k_1 c_0$, there are additional fixed points on a line $r + s = c_0 - \lambda/k_1$. This line corresponds to the one $r + s = c_0$ in §2 shifted by the back reaction $\lambda$. In fact the trajectory is the same as in §2 and radiates from the origin $r/s = r_0/s_0$ =constant, as shown in Fig.5(b).

**• Quadratically autocatalytic case:**

With a quadratic autocatalysis $k_2 > 0, k_0 = k_1 = 0$, the back reaction decimates the fixed line $r + s = c_0$ and produces several fixed points. One of them is a trivial fixed point at the origin $(r, s) = (0, 0)$, which is of no interest.

(1) With a small rate of back reaction, $\lambda < k_2 c_0^2/8$, there are other six fixed points: two on the $r$ axis, $\mathrm{U}_1 :\ (A_-, 0)$ and $\mathrm{S}_1 :\ (A_+, 0)$, two on the $s$ axis, $\mathrm{U}_2 :\ (0, A_-)$ and $\mathrm{S}_2 :\ (0, A_+)$, and two symmetric ones, $\mathrm{U}_3 :\ (B_-, B_-)$ and $\mathrm{U}_4 :\ (B_+, B_+)$, with

$$A_\pm = \frac{c_0}{2}\left(1 \pm \sqrt{1 - \frac{4\lambda}{k_2 c_0^2}}\right), \qquad B_\pm = \frac{c_0}{4}\left(1 \pm \sqrt{1 - \frac{8\lambda}{k_2 c_0^2}}\right).$$

As an illustration shown in Fig.5(c), we use parameter values $k_2 c_0^2 = 1,\ \lambda = 0.1$, which gives fixed points $\mathrm{U}_{1-4}$ and $\mathrm{S}_{1,2}$ specified by the values, $A_+ = 0.8873 c_0 =$

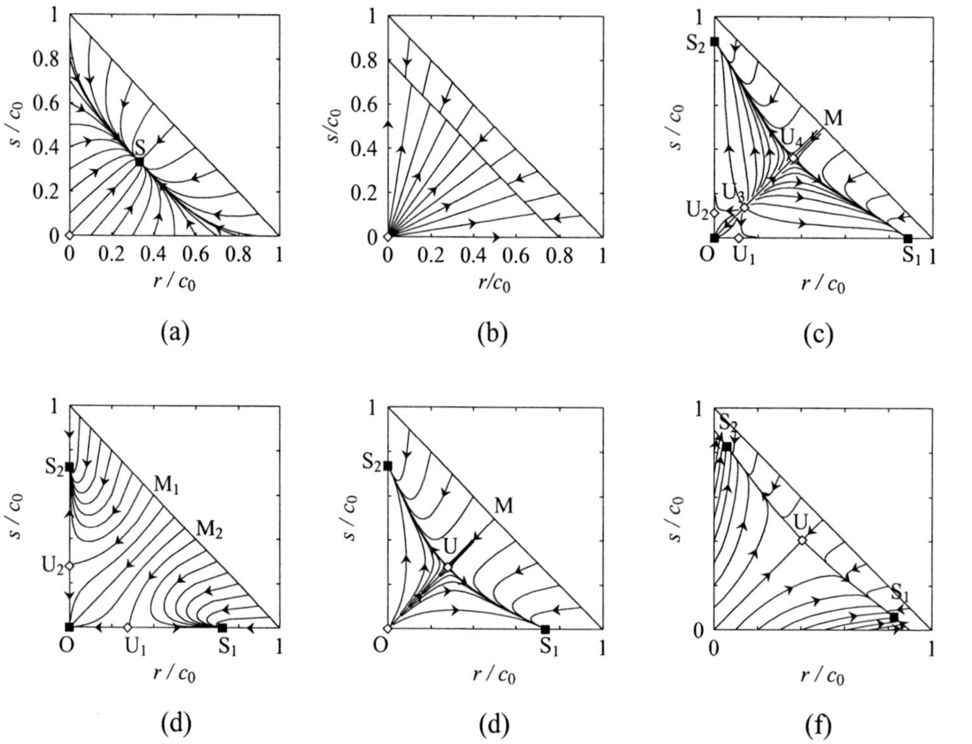

Figure 5: Flow diagrams for $(k_0,\ k_1c_0,\ k_2c_0^2,\ \lambda) =$ (a) (1, 0, 0, 1), (b) (0, 1, 0, 0.1), (c) (0, 0, 1, 0.1), (d) (0, 0, 1, 0.2), (e) (0, 0.4, 1, 0.3), and (f) (0.05, 0, 1, 0.1).

$c_0 - A_-$, and $B_+ = 0.3618c_0 = c_0/2 - B_-$. The stable fixed points $S_1$ and $S_2$ correspond to the states with complete homochirality with $\phi = \pm 1$. Except the region around the trivial fixed point O, the flow trajectory is attracted to one of the two homochiral states, $S_1$ or $S_2$. Generally, the region of initial points which will eventually be attracted to a certain fixed point is called the basin of attraction. In the present case, the $r - s$ phase space is divided into three basins of attractions to $S_1$, $S_2$, and O, by separatrixes $U_3U_1$, $U_3U_2$, and $U_3M$, where M is at $(c_0/2,\ c_0/2)$.

The time evolution of the order parameter $\phi$ is depicted in Fig.6(a) with the parameters $k_0 = k_1 = 0, k_2c_0^2 = 1, \lambda c_0 = 0.1$, the same ones used to obtain the flow diagram in Fig.5(c). The initial condition is $r_0/c_0 = 0.2 + 0.001, s_0/c_0 = 0.2$. The flow is almost symmetric as $r \approx s$ at first and approaches to the symmetric fixed point $U_4$, but since $U_4$ is unstable $r$ starts to increase at the cost of $s$ and eventually $\phi$ approaches the full value, unity.

(2) As the back reaction rate increases, two symmetric fixed points $U_3$ and $U_4$ merge to disappear for $\lambda \geq k_2c_0^2/4$. There remains two homochiral fixed points $S_1$ and $S_2$. Their basins of attraction are close to the $r$ or $s$ axes. Close to the diagonal $r = s$ the

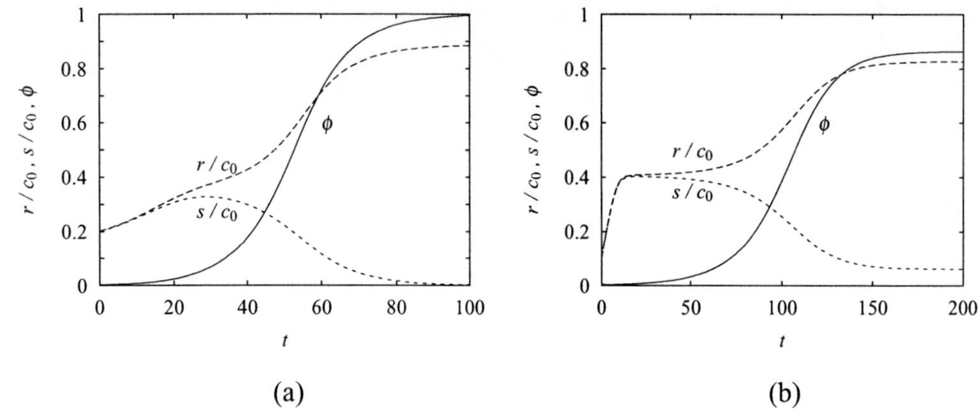


Figure 6: Time evolution of the ee $\phi$ with nonlinear autocatalysis and a back reaction (a) without random creation $k_0 = 0$, and (b) with a finte $k_0 = 0.05$. Other parameters are the same; $k_1 = 0,\ k_2 c_0^2 = 1,\ \lambda = 0.1$.

flow runs to the trivial fixed point O. The separatrixes of three basins of attractions are $U_1M_1$ and $U_2M_2$. With the parameter values $k_2 c_0^2 = 1,\ \lambda = 0.2$, fixed points $S_1$ and $S_2$ are determined with parameters $A_+ = 0.7236 c_0$ and $A_- = c_0 - A_+$, as shown in Fig.5(d).

(3) Finally, when the back reaction is too strong as $\lambda \geq k_2 c_0^2/4$, all the flow runs to the origin.

**• Linear and quadratic autocatalysis:**

We consider now the effect of a linear autocatalytic process in addition to the quadratic autocatalysis $k_2$. As far as $k_1$ is smaller than $\lambda/c_0$, the origin $r = s = 0$ is attractive, and the schematics of the flow in the $r - s$ phase space are the same with those for $k_1 = 0$. When $k_1$ is larger than $\lambda/c_0$, the origin $r = s = 0$ loses the stability, and there are only three fixed points: two stable ones on $r$ and $s$ axes $S_1$: $(C, 0)$ and $S_2$: $(0, C)$ and a symmetric and unstable one at U: $(D, D)$ with

$$C = \frac{k_2 c_0 - k_1 + \sqrt{(k_2 c_0 + k_1)^2 - 4k_2\lambda}}{2k_2}$$

$$D = \frac{k_2 c_0 - 2k_1 + \sqrt{(k_2 c_0 + 2k_1)^2 - 8k_2\lambda}}{4k_2}.$$

The typical flow in the case with $k_0 = 0, k_1 c_0 = 0.4, k_2 c_0^2 = 1, \lambda = 0.3$ is shown in Fig.5(e), which looks similar to the case with the cross inhibition shown in Fig. 3(b): complete homochiral states, $\phi = \pm 1$, is attained asymptotically even though not all the A material is consumed.

**• Autocatalysis with nonautocatalytic production**

We found that the quadratic autocatalysis with back reaction permits to attain the homochiral situation. However, in order to actuate the autocatalysis, one has to provide initial C molecules such that the initial condition lies outside the basin of attractive to the origin $r = s = 0$. The initial state outside this basin may be prepared by some outside settings due to the human factors or other reasons. Or more naturally, the reaction system involves nonautocatalytic production with a finite $k_0$. We now consider cases when the random creation of racemic mixtures by $k_0$ coexist with the autocatalytic production as well as the back reaction.

(1) With a linear autocatalysis, $k_0 > 0,\ k_1 > 0, k_2 = 0$, and $\lambda > 0$, the fixed line for $k_0 = 0$ in Fig.5(b) is replaced by a symmetric fixed point at

$$r = s = \frac{1}{4k_1}\left\{k_1c_0 - 2k_0 - \lambda + \sqrt{(k_1c_0 - 2k_0 - \lambda)^2 + 8k_0k_1c_0)}\right\} \tag{16}$$

which is stable and $\phi = 0$. Racemization effect by $k_0$ dominates and returns the system into the symmetric state, i.e. the racemic mixture.

(2) With a quadratic autocatalysis, $k_0 > 0,\ k_2 > 0,\ k_1 = 0$ and $\lambda > 0$, the situation is rather complex. Since $f(0) \neq 0$, all the fixed points stay within the triangle, defined by eq (2); $a,\ r,\ s \neq 0$. Then the fixed points satisfy the relation

$$\frac{k_0 + k_2r^2}{r} = \frac{\lambda}{a} = \frac{k_0 + k_2s^2}{s} \quad \rightarrow \quad (r-s)(k_0 - k_2rs) = 0 \tag{17}$$

The symmetry broken states stay on a hyperbola $rs = k_0/k_2$. They satisfy further the relation

$$\Lambda_a(r) \equiv k_2\left[\frac{c_0^2}{4} - \left(r - \frac{c_0}{2} + \frac{k_0}{k_2r}\right)^2\right] = \lambda > 0. \tag{18}$$

If $k_0$ is smaller than $k_2c_0^2/16$, the function $\Lambda_a(r)$ in eq.(18) has two peaks with a height $\lambda_{max} = k_2c_0^2/4$. The case with $k_0 = 0.01k_2c_0^2$ is shown in Fig.7(a). $\Lambda_a(r)$ also has a local minimum $\lambda_{min} = 2\sqrt{k_0}(c_0\sqrt{k_2} - 2\sqrt{k_0})$ at $r = \sqrt{k_0/k_2}$. Then the number of asymmetric fixed points varies as follows: For small $\lambda < \lambda_{min}$ there are two which are stable. For intermediate case with $\lambda_{min} < \lambda < \lambda_{max}$ there are four, and two among them are stable and two other are unstable. For large $\lambda > \lambda_{max}$ there is none.

If $k_0$ lies between $k_2c_0^2/16$ and $k_2c_0^2/4$, $\Lambda_a(r)$ has only a single peak with a value $\lambda'_{max} = 2\sqrt{k_0}(c_0\sqrt{k_2} - 2\sqrt{k_0}) > 0$, as in the case of $k_0 = 0.1k_2c_0^2$ shown in Fig.7(a). Therefore the number of solutions are two for small $\lambda < \lambda'_{max}$ which are stable, and none for large $\lambda > \lambda'_{max}$.

Further larger $k_0 > k_2c_0^2/4$, there is no asymmetric fixed point for any $\lambda$. Racemization effect by $k_0$ is too strong to break chiral symmetry, as is already described in relation to the time evolution of $\phi$ given in Eq.(7).

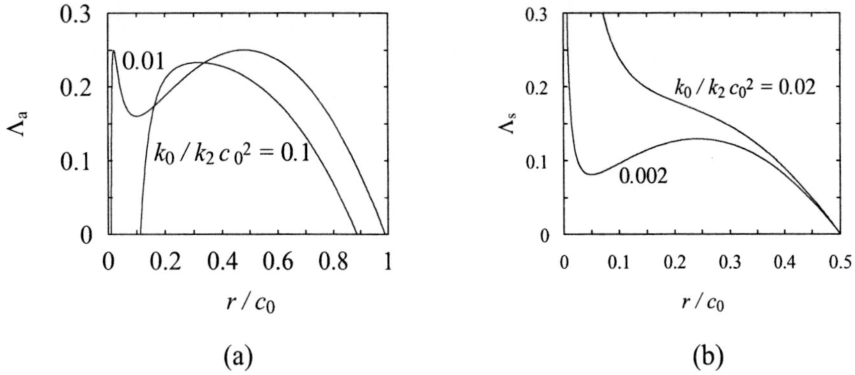


Figure 7: Functions (a) $\Lambda_a(r)$ and (b) $\Lambda_s(r)$ to determine (a) asymmetric and (b) symmetric fixed points.

As for the symmetric solutions with $r = s$, the effect of non-autocatalytic reaction $k_0$ is more effective. Since the fixed points have to satisfy a relation

$$\Lambda_s(r) \equiv \frac{k_2 c_0^3}{r}\Big[-2\Big(\frac{r}{c_0}-\frac{1}{6}\Big)^3+\Big(\frac{k_0}{k_2 c_0^2}-\frac{1}{108}\Big)\Big]+2\Big(\frac{k_2 c_0^2}{12}-k_0\Big)=\lambda, \quad (19)$$

the number of fixed points reduces to one which is unstable if $k_0$ is larger than $k_2 c_0^2/108$. The case with $k_0 = 0.02 k_2 c_0^2$ is shown in Fig.7(b). For a small rate $k_0 \leq k_2 c_0^2/108$, the numbers of symmetric fixed points depends on $\lambda$, as shown in the case with $k_0 = 0.002 k_2 c_0^2$ in Fig.7(b). For a small $\lambda$, there is one unstable, for an intermediate value of $\lambda$ there are three among which one is stable and two unstable, and for a large $\lambda$ there remains only one stable one. For example, by adding $k_0$ as small as $k_0 = 0.05$(but greater than $k_2 c_0^2/108$) to the situation shown in Fig.5(c), the origin loses the stability and there remains only a single symmetric fixed point which is unstable. Therefore, the flow diagram reduces to the one shown in Fig.5(f). The achieved ee is not complete, but is very large if $k_0$ is small, as shown in Fig.6(b). There the initial condition is chosen as $r_0/c_0 = 0.1 + 0.001$ and $s_0/c_0 = 0.1$, but the final $\phi$ does not depend on the initial values of $r_0$ or $s_0$.

# 5 Summary and Discussions

We studied various autocatalytic models with recycling processes via the back reaction or the cross inhibition, and looked for the conditions to achieve the chirality selection in a closed system. The results are summarized in Table 1.

If the production of chiral molecules R and S from the achiral material A involves a linear autocatalysis, the cross inhibition between R and S is necessary to realise the ho-

Table 1: Classification of reaction models with parameters defined in the bottom column. Symbol $\geq$ means that the value is non-negative and $>$ means that the value must be positive. B means that the final state is chiral symmetry breaking and S means otherwise. Details for each entry are explained in the indicated sections.

| | $k_0$ | $k_1$ | $k_2$ | $\lambda$ | $\mu$ | SB | section |
|---|---|---|---|---|---|---|---|
| (1) | $\geq$ | $\geq$ | $\geq$ | 0 | 0 | S | §2 |
| (2) | $\geq$ | 0 | 0 | 0 | $>$ | S | §3 |
| (3) | $\geq$ | $>$ | $\geq$ | 0 | $>$ | B | §3 |
| (4) | $\geq$ | $\geq$ | 0 | $>$ | 0 | S | §4 |
| (5) | $\geq$ | $\geq$ | $>$ | $>$ | 0 | B | §4 |

$$\dot{r} = (k_0 + k_1 r + k_2 r^2)a - \lambda r - \mu rs$$
$$\dot{s} = (k_0 + k_1 s + k_2 s^2)a - \lambda s - \mu rs$$
$$a = c_0 - r - s$$

mochirality, as is summarized for the cases (1)-(4) in Table 1. The case (3) corresponds to an extension of Frank's model to a closed chemical system.

With a quadratic autocatalysis but without any recycling processes as in the case (1), only the amplification of enantiomeric excess (ee) is achieved, but the final value of ee is arbitrary, depending on the details of the initial values of chiral as well as achiral materials' concentrations. Therefore we do not regard the case (1) as the chiral symmetry breaking in a strict sense.

With a linear back reaction process and a quadratic autocatalysis as a special case of (5), the complete homochirality is possible when the rate of back reaction is sufficiently small. The minority enantiomer is decomposed back to the material A and recycled to produce the majority enantiomer and vice versa. This decomposition process works more effectively for the minority because of the suppressed production due to the nonlinear autoctalysis.

However, with only autocatalytic processes, the initial embryos of chiral molecules are required. The initial state outside the basin of attractive racemic state may be prepared by some outside settings due to the human factors or other reasons. Or more naturally, the reaction system involves nonautocatalytic production with a finite $k_0$. For a small value of $k_0$, the phase space structure of the flow and fixed points is not qualitatively affected and resembles with that in the absence of $k_0$. However, for $k_0$ between $k_2c_0^2/108$ and $k_2c_0^2/4$, this process affects the structure of racemic fixed points so drastically that stable racemic state disappears, as shown in Fig.5(f).

We have studied two types of simple models (3) and (5) where the chiral symmetry can be broken in the closed system, and it is found that complete or almost complete homochiral states can be achieved. One type of the model with the cross inhibition (3) seems powerful for symmetry breaking, but the chemical system which obeys the process has not been found so far and it is very interesting to find actual chemical system. On the other hand, the nonlinear autocatalytic system with the ee amplification is found by Soai's group and thus additional back reaction could provide the chirality selection. But we are not certain whether there is a system with a back reaction which proceeds in a measurerable time for the experiment.

Based on the analysis of the present simple models in a closed system, we would like to venture to speculate a possible scenario for the initial preparation of constituent materials of life as follows. In a confined nook and cranny somewhere in the premordial sea, there was a very favorable conditions for the autocatalytic reactions to take place, and after a sufficiently long time the initial homochiral selection was established. Once this selection was completed in this tiny closed corner, the chiral molecules began to spread out and to conquer eventually the surrounding region and futher all over the earth.

So far, we have assumed that the system is homogenious, such that the whole system is described by only a space-independent concentrations. In the actual situations, the concentrations should vary in space. We have introduced a lattice model for the reaction-diffusion system and studied the spreading of chiral symmetry breaking. The interested readers are invited to refer to the original paper[15].

### Acknowledgements

The work is supported by the Gakuji Shinkou Shikin, Keio University.

## Note Added in Proof

After the submission of the present paper, chirality selection in a closed system was proven experimentally in the domain of crystal growth [16], and we extended our closed system model to the crystal growth problem, and showed that the recycling effect is essential in the chirality selection of the crystal growth as well [17]. Also in another work [18] we proposed that the complete chirality selection in Soai reaction is possible if achiral substrate flows in and everything flows out in a steady flow. This model was shown to have mathematically similar structures as the model in a closed system. There [18], we also dealt with the polymerization process in a steady flow system, and found that the catalytic effect of the polymer breaks the chiral symmetry.

## References

[1] W. A. Bonner: *Topics Stereochem.* **18** (1988) 1.

[2] B. L. Feringa and R. A. van Delden: *Angew. Chem. Int. Ed.* **38** (1999) 3418.

[3] S. F. Mason and G. E. Tranter: *Proc. R. Soc. Lond.* **A 397** (1985) 45.

[4] D. K. Kondepudi and G. W. Nelson: *Nature* **314** (1985) 438.

[5] W. J. Meiring: *Nature* **329** (1987) 712.

[6] J. L. Bada: *Nature* **374** (1995) 594.

[7] J. Bailey, A. Chrysostomou, J. H. Hough, T. M. Gledhill, A. McCall, S. Clark, F. Ménard and M. Tamura: *Science* **281** (1998) 672.

[8] R. M. Hazen, T. R. Filley and G. A. Goodfriend: *Proc. Natl. Acad. Sci.* **98** (2001) 5487.

[9] F. C. Frank: *Biochimi. Biophys. Acta* **11** (1953) 459.

[10] K. Soai, S. Niwa and H. Hori: *J. Chem. Soc. Chem. Commun.* (1990) 982.

[11] K. Soai, T. Shibata, H. Morioka and K. Choji: *Nature* **378** (1995) 767.

[12] I. Sato, D. Omiya, K. Tsukiyama, Y. Ogi and K. Soai: *Tetrahedron Asymmetry* **12** (2001) 1965.

[13] I. Sato, D. Omiya, H. Igarashi, K. Kato, Y. Ogi, K. Tsukiyama and K. Soai: *Tetrahedron Asymmetry* **14** (2003) 975.

[14] Y. Saito and H. Hyuga, *J. Phys. Soc. Jpn.* **73** (2004) 33.

[15] Y. Saito and H. Hyuga, *J. Phys. Soc. Jpn.* **73** (2004) 1685.

[16] C. Viedma, *Phys. Rev. Lett.* **94** (2005) 065504.

[17] Y. Saito and H. Hyuga, *J. Phys. Soc. Jpn.* **74** (2005) 535.

[18] Y. Saito and H. Hyuga, *J. Phys. Soc. Jpn.* **74** (2005) 1629.

In: Progress in Chemical Physics Research
Editor: A. N. Linke, pp. 81-105
ISBN 1-59454-451-4


*Chapter 4*

# AB INITIO STUDIES OF LIQUID AND AMORPHOUS GE

***Jeng-Da Chai****
Institute for Physical Science and Technology, University of Maryland,
College Park, Maryland 20742
***D. Stroud***†
Department of Physics, The Ohio State University,
Columbus, Ohio 43210

**Abstract**

We review our previous work on the dynamic structure factor $S(k,\omega)$ of liquid Ge ($\ell$-Ge) at temperature $T = 1250$ K, and of amorphous Ge ($a$-Ge) at $T = 300$ K, using *ab initio* molecular dynamics [Phys. Rev. **B67**, 104205 (2003)]. The electronic energy is computed using density-functional theory, primarily in the generalized gradient approximation, together with a plane wave representation of the wave functions and ultra-soft pseudopotentials. We use a 64-atom cell with periodic boundary conditions, and calculate averages over runs of up to about 16 ps. The calculated liquid $S(k,\omega)$ agrees qualitatively with that obtained by Hosokawa *et al*, using inelastic X-ray scattering. In $a$-Ge, we find that the calculated $S(k,\omega)$ is in qualitative agreement with that obtained experimentally by Maley *et al*. Our results suggest that the *ab initio* approach is sufficient to allow approximate calculations of $S(k,\omega)$ in both liquid and amorphous materials.

## 1 Introduction

Ge is a well-known semiconductor in its solid phase, but becomes metallic in its liquid phase. Liquid Ge ($\ell$-Ge) has, near its melting point, an electrical conductivity characteristic

*E-mail address: jdchai@wam.umd.edu
†E-mail address:stroud@mps.ohio-state.edu

of a reasonably good metal ($\sim 1.6 \times 10^{-4}\Omega^{-1}\text{cm}^{-1}$[1]), but it retains some residual structural features of the solid semiconductor. For example, the static structure factor $S(k)$, has a shoulder on the high-$k$ side of its first (principal) peak, which is believed to be due to a residual tetrahedral short-range order. This shoulder is absent in more conventional liquid metals such as Na or Al, which have a more close-packed structure in the liquid state and a shoulderless first peak in the structure factor. Similarly, the bond-angle distribution function just above melting is believed to have peaks at *two* angles, one near $60^o$ and characteristic of close packing, and one near $108^o$, indicative of tetrahedral short range order. This latter peak rapidly disappears with increasing temperature in the liquid state. These striking properties of $\ell$-Ge have been studied theoretically by several groups. Their methods fall into two broad classes: empirical and first-principles. A typical empirical calculation is that of Yu *et al*[2], who calculate the structural properties of $\ell$-Ge assuming that the interatomic potentials in $\ell$-Ge are a sum of two-body and three-body potentials of the form proposed by Stillinger and Weber[3]. These authors find, in agreement with experiment, that there is a high-k shoulder on the first peak of S(k) just above melting, which fades away with increasing temperature. However, since in this model *all* the potential energy is described by a sum of two-body and three-body interactions, the interatomic forces are probably stronger, and the ionic diffusion coefficient is correspondingly smaller, than their actual values.

In the second approach, the electronic degrees of freedom are taken explicitly into account. If the electron-ion interaction is sufficiently weak, it can be treated by linear response theory[4]. In linear response, the total energy in a given ionic configuration is a term which is independent of the ionic arrangement, plus a sum of two-body ion-ion effective interactions. These interactions typically do not give the bond-angle-dependent forces which are present in the experiments, unless the calculations are carried to third order in the electron-ion pseudopotential[4], or unless electronic fluctuation forces are included[5] Such interactions are, however, included in the so-called *ab initio* approach, in which the forces on the ions are calculated from first principles, using the Hellman-Feynman theorem together with density-functional theory[6, 7, 8] to treat the energy of the inhomogeneous electron gas. This approach not only correctly gives the bond-angle-dependent ion-ion interactions, but also, when combined with standard molecular dynamics techniques, provides a good account of the electronic properties and such dynamical ionic properties as the ionic self-diffusion coefficients.

This combined approach, usually known as *ab initio* molecular dynamics, was pioneered by Car and Parrinello[9], and, in somewhat different form, has been applied to a wide range of liquid metals and alloys, including $\ell$-Ge[10, 11, 12], $\ell$-Ga$_x$Ge$_{1-x}$[13], stoichiometric III-V materials such as $\ell$-GaAs, $\ell$-GaP, and $\ell$-InP[14, 15], nonstoichiometric $\ell$-Ga$_x$As$_{1-x}$[16], $\ell$-CdTe[17], and $\ell$-ZnTe[18], among other materials which are semiconducting in their solid phases. It has been employed to calculate a wide range of properties of these materials, including the static structure factor, bond-angle distribution function, single-particle electronic density of states, d. c. and a. c. electrical conductivity, and ionic self-diffusion coefficient. The calculations generally agree quite well with available experiment.

A similar *ab initio* approach has also been applied extensively to a variety of amorphous semiconductors, usually obtained by quenching an equilibrated liquid state from the melt. For example, Car, Parrinello and their collaborators have used their own *ab initio* approach (based on treating the Fourier components of the electronic wave functions as fictitious classical variables) to obtain many structural and electronic properties of amorphous Si[19, 20]. A similar approach has been used by Lee and Chang[21]. Kresse and Hafner[10] obtained both S(k) and g(r), as well as many electronic properties, of $a$-Ge, using an *ab initio* approach similar to that used here, in which the forces are obtained directly from the Hellmann-Feynman theorem and no use is made of fictitious dynamical variables for the electrons, as in the Car-Parrinello approach. A similar calculation for $a$-Si has been carried out by Cooper *et al*[22], also making use of a plane wave basis and treating the electron density functional in the generalized gradient approximation (GGA)[23]. More recently, a number of calculations for $a$-Si and other amorphous semiconductors have been carried out by Sankey *et al*[24], and by Drabold and collaborators [25]. These calculations use *ab initio* molecular dynamics and electronic density functional theory, but in a localized basis. A recent study, in which S(k) and g(r) were computed for several *ab initio* structural models of $a$-Si, has been carried out by Alvarez *et al*[26].

Finally, we mention a third approach, intermediate between empirical and *ab initio* molecular dynamics, and generally known as tight-binding molecular dynamics. In this approach, the electronic part of the total energy is described using a general tight-binding Hamiltonian for the band electrons. The hopping matrix elements depend on separation between the ions, and additional terms are included to account for the various Coulomb energies in a consistent way. The parameters can be fitted to *ab initio* calculations, and forces on the ions can be derived from the separation-dependence of the hopping matrix elements. This approach has been used, e. g., to treat $\ell$-Si[27], $a$-Si[28], and liquid compound semiconductors such as $\ell$-GaAs and $\ell$-GaSb[29]. Results are in quite good agreement with experiment.

In this paper, we review the application of *ab initio* molecular dynamics to another dynamical property of the ions: the *dynamical structure factor*, denoted $S(k,\omega)$. While no fundamentally new theory is required to calculate $S(k,\omega)$, this quantity provides additional information about the time-dependent ionic response beyond what can be extracted from other quantities [30]. The work we review is the first to calculate $S(k,\omega)$ using *ab initio* molecular dynamics. Here, we will describe calculations of $S(k,\omega)$ for $\ell$-Ge, where some recent experiments[31] provide data for comparison, and also for amorphous Ge ($a$-Ge). In the latter case, using a series of approximations described below, we are able to infer the *vibrational* density of states of as-quenched $a$-Ge near temperature $T = 300$K in reasonable agreement with experiment. The calculated $S(k,\omega)$ for the liquid also agrees quite well with experiment, especially considering the computational uncertainties inherent in an *ab initio* simulation with its necessarily small number of atoms and limited time intervals.

The remainder of this paper is organized as follows. A brief review of the theory and calculational method is given in Section II and III respectively. The results are reviewed in Section IV, followed by a discussion in Section V.

## 2 Finite-Temperature Density-Functional Theory

The original Hohenberg-Kohn theorem of density-functional theory (DFT) [6] was generalized to finite temperatures $T_{el}$ by Mermin [8]. In this section, we give a brief review of the finite-temperature density-functional theory, in a form where it can be used in ab initio molecular dynamics calculations. We use atomic units here and throughout this article.

In the Born-Oppenheimer approximation, the total potential energy U of a system of N classical ions, enclosed in a volume V, and interacting with $N_{el}$=NZ valence electrons through an electron-ion pseudopotential $v_{ei}(r)$, is a functional of the valence electron density $\rho(\mathbf{r})$, and also a function of the ion positions $\mathbf{R}_l$ (l=1,2,...,N). This potential energy can be expressed as the sum of a direct Coulombic interaction between the ions and the free energy $F_{el}$ of the electronic system subject to the external potential created by the ions, which we denote $V_{ext}(\mathbf{r}, \{\mathbf{R}_l\}) = \sum_{l=1}^{N} v_{ei}(|\mathbf{r} - \mathbf{R}_l|)$. This sum may be written

$$U[\rho(\mathbf{r}), \{\mathbf{R}_l\}] = F_{el}[\rho(\mathbf{r})] + \sum_{l<l'} W(\mathbf{R}_l - \mathbf{R}_{l'}), \tag{1}$$

where $W(\mathbf{R} - \mathbf{R}') = Z^2/|\mathbf{R} - \mathbf{R}'|$ is the Coulomb interaction between ions of charge $Z$ located at $\mathbf{R}$ and $\mathbf{R}'$.

The free energy functional of the electronic system $F_{el}[\rho(\mathbf{r})]$ can be partitioned into various terms as follows:[32, 33]

$$F_{el}[\rho(\mathbf{r})] = A_s[\rho(\mathbf{r})] + \int \rho(\mathbf{r}) V_{ext}(\mathbf{r}) d\mathbf{r} + \frac{1}{2} \int \frac{\rho(\mathbf{r})\rho(\mathbf{r}')}{|\mathbf{r} - \mathbf{r}'|} d\mathbf{r} d\mathbf{r}' + F_{xc}[\rho(\mathbf{r})]. \tag{2}$$

Here the different terms are $A_s[\rho(\mathbf{r})]$, the Helmholz free energy functional for a reference system of noninteracting electrons; the interaction energy between the valence electrons and the external potential provided by the instantaneous ionic configuration; the classical electron-electron potential energy; and the exchange-correlation correction $F_{xc}[\rho(\mathbf{r})]$ to the free energy. $A_s[\rho(\mathbf{r})]$ has contributions from both the kinetic energy $T_s[\rho(\mathbf{r})]$ and the entropy $S_s[\rho(\mathbf{r})]$ of the reference system, and may be written

$$A_s[\rho(\mathbf{r})] = T_s[\rho(\mathbf{r})] - T_{el} S_s[\rho(\mathbf{r})], \tag{3}$$

where $T_{el}$ is the temperature of the electronic system.

For a given set of ionic positions, the equilibrium electronic number density $\rho(\mathbf{r})$ is obtained from the following variational principle:

$$\frac{\delta}{\delta \rho(\mathbf{r})} (U[\rho(\mathbf{r}), \{\mathbf{R}_l\}] - \mu \int \rho(\mathbf{r}) d\mathbf{r}) = 0. \tag{4}$$

Here $\mu$ is the electron chemical potential, which is chosen to give the desired number of valence electrons $N_{el}$. The Euler equation associated with eq. (4) is

$$\mu = V_{A_s}(\mathbf{r}; [\rho]) + V_{eff}(\mathbf{r}; [\rho]). \tag{5}$$

Here

$$V_{A_s}(\mathbf{r};[\rho]) = \frac{\delta A_s[\rho]}{\delta\rho(\mathbf{r})} \tag{6}$$

is a functional derivative of $A_s[\rho]$, and is the "Helmholz free energy potential" (or the kinetic potential for $T_{el} = 0$) [34]. $V_{eff}(\mathbf{r};[\rho])$ is an effective one-body potential, given by

$$V_{eff}(\mathbf{r};[\rho]) = V_{ext}(\mathbf{r}) + \int \frac{\rho(\mathbf{r}')}{|\mathbf{r}-\mathbf{r}'|}d\mathbf{r}' + \frac{\delta F_{xc}[\rho]}{\delta\rho(\mathbf{r})}. \tag{7}$$

It is usual to make several approximations for the functionals $A_s[\rho]$ and $F_{xc}[\rho]$, and the function $V_{ext}(\mathbf{r})$. First, $F_{xc}[\rho]$ is only a minor contribution to $F_{el}[\rho]$. As a result, this functional is often calculated using the local density approximation (LDA) and or the generalized gradient approximation (GGA); both are widely used in many applications [32].

Furthermore, $F_{xc}$ differs negligibly from its zero-temperature value, the exchange-correlation energy $E_{xc}$, provided $T_{el} < 0.15T_f$, where $T_f$ is the Fermi temperature [35]. Most applications involve temperatures satisfying this inequality. The external field $V_{ext}(\mathbf{r})$ contains the electron-ion pseudopotential $v_{ei}(r)$, on which other approximations are made. For example, it is common to approximate $v_{ei}$ as local and energy-independent.

Once an accurately approximated $A_s[\rho]$ is available, one can, in principle, obtain $V_{A_s}(\mathbf{r};[\rho])$ by functional differentiation, and from this functional, solve eq. (5) to obtain the equilibrium electronic density. However, this approach has proved to be very challenging in practice.

Typically, the kinetic energy $T_s[\rho]$ is much larger than the entropy contribution $T_{el}S_s[\rho]$ in the range of temperature interested. Therefore, $A_s[\rho]$ is generally well approximated by $T_s[\rho]$. However, simple local density approximations for $T_s[\rho]$ based on the Thomas-Fermi (TF) theory are not very accurate, and make several qualitatively incorrect predictions [36, 37]. More recent developments based on so-called "orbital-free"(OF) approximations for the kinetic energy functional $T_s[\rho]$ [38, 39, 40, 41], and the kinetic potential $V_{T_s}(\mathbf{r};[\rho])$ [42, 43] are still insufficient to make quantitative predictions. On the other hand, although these OF methods are currently unable to make accurate predictions, they are computationally appealing, because the computational time required to use them is independent of the number of valence electrons. Therefore, further refinement of such OF methods in DFT may lead to valuable new approaches for investigating the properties of large systems, especially metallic systems.

Instead of using approximate forms for $A_s[\rho]$ or $V_{A_s}(\mathbf{r};[\rho])$, Kohn and Sham suggested a different way to compute the numerical value of $A_s$ *exactly*, by introducing a set of $N_{el}$ orbitals satisfying the coupled KS equations that describe the model system [7]. Using these orbitals one can determine both $A_s$ and the valence electron density $\rho(\mathbf{r})$. The orbitals $\psi_i$ satisfy the Schrödinger-like equation

$$\left[-\frac{1}{2}\nabla^2 + V_{eff}(\mathbf{r})\right]\psi_i = \epsilon_i\psi_i, \tag{8}$$

and are chosen to be orthonormal. The set of equations is made self-consistent by requiring that

$$\rho(\mathbf{r}) = \sum_i f_i |\psi_i(\mathbf{r}).|^2 \tag{9}$$

Here $f_i$ is the Fermi occupation number, given by

$$f_i = f(\epsilon_i - \mu) = (exp(\frac{\epsilon_i - \mu}{k_B T_{el}}) + 1)^{-1}. \tag{10}$$

The kinetic energy $T_s$ is given by

$$T_s[\{\psi_i(\mathbf{r})\}, \{f_i\}] = \sum_i f_i \int \psi_i(-\frac{1}{2}\nabla^2)\psi_i d\mathbf{r} = \sum_i \epsilon_i f_i - \int \rho(\mathbf{r}) V_{eff}(\mathbf{r}) d\mathbf{r}, \tag{11}$$

and the electronic entropy $S_s$ by

$$S_s[\{f_i\}] = -2k_B \sum_i [f_i ln f_i + (1 - f_i) ln(1 - f_i)] \tag{12}$$

Eqs. (7,8,9,10) have to be solved self-consistently. The resulting valence electron density can be used to computed the electronic free energy $F_{el}[\rho]$; the electronic kinetic energy $T_s$ and entropy $S_s$ are given by eq(11) and eq(12) [32].

The use of finite-temperature DFT is important, in order to eliminate the instability of the KS equations at $T_{el} = 0$. In practice, $T_{el}$ is usually regarded as a fictitious temperature (chosen for maintaining the stability), and does not necessarily have to equal the ionic temperature T. Furthermore, the broadening function in eq. (10) need not be the Fermi function, but may be chosen to be some other convenient function. Several possible broadening functions have been proposed [10, 44].

The forces on the individual ions can now be obtained using the Hellmann-Feynman theorem. According to this theorem, the force $\mathbf{F}_I$ acting on the $I^{th}$ ion is found by differentiating the potential energy of the system with respect to the ionic coordinate $\mathbf{R}_I$[45, 46]. Specifically,

$$\mathbf{F}_I = -\frac{\partial U}{\partial \mathbf{R}_I} = -\sum_i f_i < \psi_i | \frac{\partial F_{el}[\rho(\mathbf{r})]}{\partial \mathbf{R}_I} | \psi_i > -\frac{\partial}{\partial \mathbf{R}_I} \sum_{l<l'} W(\mathbf{R}_l - \mathbf{R}_{l'}) \tag{13}$$

Given the force, one simply applies Newton's second law to get the equations of motion of the $N$ classical ions:

$$M_I \ddot{\mathbf{R}}_I = \mathbf{F}_I \tag{14}$$

In practice, the ions are treated classically, even though the origin of the force is very much quantum-mechanical.

In solving these equations of motion, it is important to maintain the ionic temperature T at a fixed value. This is typically accomplished by rescaling the total kinetic energy of the ions $K_I = \sum_{l=1}^{N} \frac{\mathbf{P}_l^2}{2M_l}$ to satisfy

$$T = \frac{2K_I}{3k_B(N-1)} \tag{15}$$

every few time steps. The factor of N-1, not N, appears here so as to conserve the total momentum of the system of ions.

In summary, for each instantaneous configuration of ions, one computes $F_{el}[\rho(\mathbf{r})]$ and hence $U[\rho(\mathbf{r}), \{\mathbf{R}_l\}]$ by solving the finite-temperature Kohn-Sham equations (7),(8),(9), and (10). The Hellmann-Feynman force $\mathbf{F}_I$ is then computed from eq. (13). By numerically solving Newton's equations of motion for the ions and rescaling the kinetic energy of the ions at every time step, one moves the ions to a new configuration. The entire process is repeated for a sufficiently large number of time steps to compute the desired property with good statistical accuracy.

# 3 Method

The method we have used to calculate the dynamic structure factor[30] is similar to that described in several previous papers for other properties of liquid semiconductors[12, 13, 16], but uses the Vienna Ab Initio Simulation Package (VASP), whose workings have been extensively described in the literature[47]." Briefly, the calculation involves two parts. First, for a given ionic configuration, the total electronic energy is calculated, using an approximate form of the Kohn-Sham free energy density functional, and the force on each ion is also calculated, using the Hellmann-Feynman theorem. Second, Newton's equations of motion are integrated numerically for the ions, using a suitable time step. The process is repeated for as many time steps as are needed to calculate the desired quantity. To hold the temperature constant, we use the canonical ensemble with the velocity rescaled at each time step. Further details of this approach are given in Ref. [12].

The VASP code uses the ultrasoft Vanderbilt pseudopotentials[48], a plane wave basis for the wave functions, with the original Monkhorst-Pack ($3 \times 3 \times 3$) k-space meshes[49] and a total of 21,952 plane waves, corresponding to an energy cut-off of 104.4eV. We have used a finite-temperature version of the Kohn-Sham theory[8] in which the electron gas Helmholtz free energy is calculated on each time step. This version also broadens the one-electron energy levels to make the k-space sums converge more rapidly. The method of Methfessel-Paxton of order 1 is used for the broadening function [44]., with the fictitious temperature for the electronic subsystem $k_B T_{el}$=0.2 eV. Most of our calculations were done using the generalized gradient approximation (GGA)[23] for the exchange-correlation energy (we use the particular form of the GGA developed by Perdew and Wang[23]), but some are also carried out using the local-density approximation (LDA).

In our iteration of Newton's Laws in liquid Ge ($\ell$-Ge), we typically started from the diamond structure (at the experimental liquid state density for the temperature of interest), then iterated for 901 time steps, each of 10 fs, using the LDA. To obtain S(k) within the GGA, we started from the LDA configuration after 601 time steps, then iterated using the GGA for an additional 1641 10-fs time steps, or 16.41 ps. We calculate the GGA S(k) by averaging over an interval of 13.41 ps within this 16.41 time interval, starting a time $t_2$ after the start of the GGA simulation. We average over all $t_2$'s from 1.0 to 3.0 ps.

For comparison, we have also calculated S(k) within the LDA. This S(k) is obtained by averaging over 601 time steps of the 901 time-step LDA simulation. This 601-step interval is chosen to start a time $t_1$ after the start of this simulation; the calculated LDA S(k) is also averaged over all $t_1$'s from 1.0 ps to 3.0 ps.

To calculate quantities for amorphous Ge (a-Ge), we start with Ge in the diamond structure at T = 1600K but at the calculated liquid density for that temperature, as given in Ref. [12]. Next, we quench this sample to 300 K, cooling at a uniform rate over, so as to reach 300 K in about 3.25 ps (in 10-fs time steps). Finally, starting from T = 300 K, we iterate for a further 897 time steps, each of 10 fs, or 8.97 ps, using the LDA. The LDA S(k) is then obtained by averaging over 5.97 of those 8.97 ps, starting a time $t_1$ after the system is reached 300K; we also average this S(k) over all $t_1$'s from 1.0 to 3.0 ps. To obtain S(k) within the GGA, we start the GGA after 5.7 ps of the LDA simulation, and then iterate using the GGA for an additional 18.11 ps in 10-fs time steps. The GGA S(k) is obtained by averaging over a 15.11 ps time interval of this 18.11 ps run, starting a time $t_2$ after the start of the GGA simulation; we also average over all values of $t_2$ from 1.0 to 3.0 ps.

The reader may be concerned that the 3.25 ps quench time is very short, very unrepresentative of a realistic quench, and very likely to produce numerous defects in the quenched structure, which are not typical of the $a$-Ge studied in experiments. In defense, we note that some of these defects may be annealed out in the subsequent relaxation at low temperatures, which is carried out before the averages are taken. In addition, the static structure factor we obtain agrees well with experiment, and the dynamic structure factor is also consistent with experiment, as discussed below. Thus, the quench procedure appears to produce a material rather similar to some of those studied experimentally. Finally, we note that experiments on $a$-Ge themselves show some variation, depending on the exact method of sample preparation.

We have used the procedure outlined above to calculate various properties of $\ell$-Ge and $a$-Ge. Most of these calculated properties have been described in previous papers, using slightly different methods, and therefore will be discussed here only very briefly[50]. However, our results for the dynamic structure factor $S(k,\omega)$ as a function of wave vector k and frequency $\omega$ are new, and will be described in detail. We also present our calculated static structure factor $S(k)$, which is needed in order to understand the dynamical results.

$S(k)$ is defined by the relation

$$S(k) = \frac{1}{N}\langle\sum_{i,j}\exp[i\mathbf{k}\cdot(\mathbf{r}_i(t)-\mathbf{r}_j(t))]\rangle_{t_0} - N\delta_{\mathbf{k},0}, \tag{16}$$

where $\mathbf{r}_i$ is the position of the i$^{th}$ ion at time t, $N$ is the number of ions in the sample, and the triangular brackets denote an average over the sampling time. In all our calculations, we have used a cubic cell with $N = 64$ and periodic boundary conditions in all three directions. This choice of particle number and cell shape is compatible with any possible diamond-structure Ge within the computational cell.

$S(\mathbf{k},\omega)$ is defined by the relation (for $\mathbf{k}\neq 0, \omega\neq 0$)

$$S(\mathbf{k},\omega) = \frac{1}{2\pi N}\int_{-\infty}^{\infty}\exp(i\omega t)\langle\rho(\mathbf{k},t)\rho(-\mathbf{k},0)\rangle dt, \tag{17}$$

where the Fourier component $\rho(\mathbf{k}, t)$ of the number density is defined by

$$\rho(\mathbf{k}, t) = \sum_{i=1}^{N} \exp(-i\mathbf{k} \cdot \mathbf{r}_i(t)). \tag{18}$$

In our calculations, the average $\langle ... \rangle$ is computed as

$$\langle \rho(\mathbf{k}, t)\rho(-\mathbf{k}, 0) \rangle = \frac{1}{\Delta t_1} \int_0^{\Delta t_1} \rho(\mathbf{k}, t_1 + t)\rho(-\mathbf{k}, t_1)dt_1 \tag{19}$$

over a suitable range of initial times $t_1$. Because of the expected isotropy of the liquid or amorphous phase, which should hold in the limit of large N, $S(\mathbf{k}, \omega)$ should be a function only of the magnitude $k$ rather than the vector $\mathbf{k}$, as should the structure factor $S(\mathbf{k})$.

Our calculations are carried out over relatively short times. To reduce statistical errors, we therefore first calculate

$$S(\mathbf{k}, \omega, t_1, t_2) = \frac{1}{\pi N} \int_0^{t_2} dt \rho(\mathbf{k}, t_1 + t)\rho(-\mathbf{k}, t_1) \exp(i\omega t). \tag{20}$$

For large enough $t_2$, $S(\mathbf{k}, \omega, t_1, t_2)$ should become independent of $t_2$ but will still retain some dependence on $t_1$. Therefore, in the liquid, we obtain our calculated dynamic structure factor, $S_{calc}(\mathbf{k}, \omega)$, by averaging over a suitable range of $t_1$ from 0 to $\Delta t_1$:

$$S_{calc}(\mathbf{k}, \omega) = \frac{1}{\Delta t_1} \int_0^{\Delta t_1} dt_1 S(\mathbf{k}, \omega, t_1, t_2). \tag{21}$$

We choose our initial time in the $t_1$ integral to be 1 ps after the start of the GGA calculation, and (in the liquid) $\Delta t_1 = 6$ ps. We choose the final MD time $t_2 = 16.41$ps. For $a$-Ge, we use the same procedure but $t_2 = 18.11$ps in our simulations. For our finite simulational sample, the calculated $S(\mathbf{k})$ and $S(\mathbf{k}, \omega)$ will, in fact, depend on the direction as well as the magnitude of $\mathbf{k}$. To suppress this finite-size effect, we average the calculated $S(\mathbf{k})$ and $S(\mathbf{k}, \omega)$ over all $\mathbf{k}$ values of the same length. This averaging considerably reduces statistical error in both $S(k, \omega)$ and $S(k)$.

Finally, we have also incorporated the experimental resolution functions into our plotted values of $S(k, \omega)$. Specifically, we generally plot $S_{av}(k, \omega)/S(k)$, where $S_{av}(k, \omega)$ is obtained from the (already orientationally averaged) $S(k, \omega)$ using the formula

$$S_{av}(k, \omega) = \int_{-\infty}^{\infty} R(\omega - \omega')S(k, \omega')d\omega', \tag{22}$$

where the resolution function $R(\omega)$ (normalized so that $\int_{-\infty}^{\infty} R(\omega)d\omega = 1$) is

$$R(\omega) = \frac{1}{\sqrt{\pi}\omega_0} \exp(-\omega^2/\omega_0^2). \tag{23}$$

In an isotropic liquid, we must have $S(k, -\omega) = S(k, \omega)$, since our ions are assumed to move classically under the calculated first-principles force. Our orientational averaging procedure guarantees that this will be satisfied identically in our calculations, since for every $\mathbf{k}$, we always include $-\mathbf{k}$ in the same average. We will nonetheless show results for negative $\omega$ for clarity, but they do not provide any additional information.

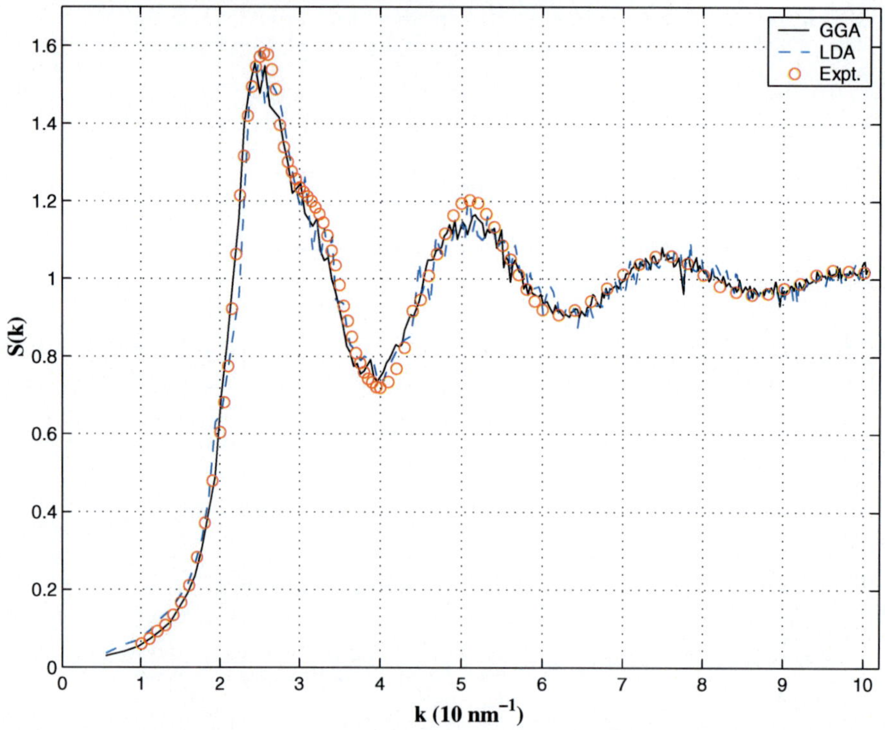


Figure 1: Static structure factor S(k) for $\ell$-Ge at $T = 1250K$, just above the experimental melting temperature. Full curve: present work, as calculated using the generalized gradient approximation (GGA; see text). Dashed curve: present work, but using the local density approximation (LDA; see text). Open circles: measured S(k) near $T = 1250$ K, as given in Ref. [51].

# 4 Results

## 4.1 S(k) for $\ell$-Ge and $a$-Ge

In Fig. 1, we show the calculated $S(k)$ for $\ell$-Ge at $T = 1250$ K, as obtained using the procedure described in Section II. The two calculated curves are obtained using the GGA and the LDA for the electronic energy-density functional; they lead to nearly identical results. The calculated $S(k)$ shows the well-known characteristics already found in previous simulations[10, 12]. Most notably, there is a shoulder on the high-$k$ side of the principal peak, which is believed to arise from residual short-range tetrahedral order persisting into the liquid phase just above melting. We also show the experimental results of Waseda *et al*[51]; agreement between simulation and experiment is good, and in particular the shoulder seen in experiment is also present in both calculated curves (as observed also in previous simulations).

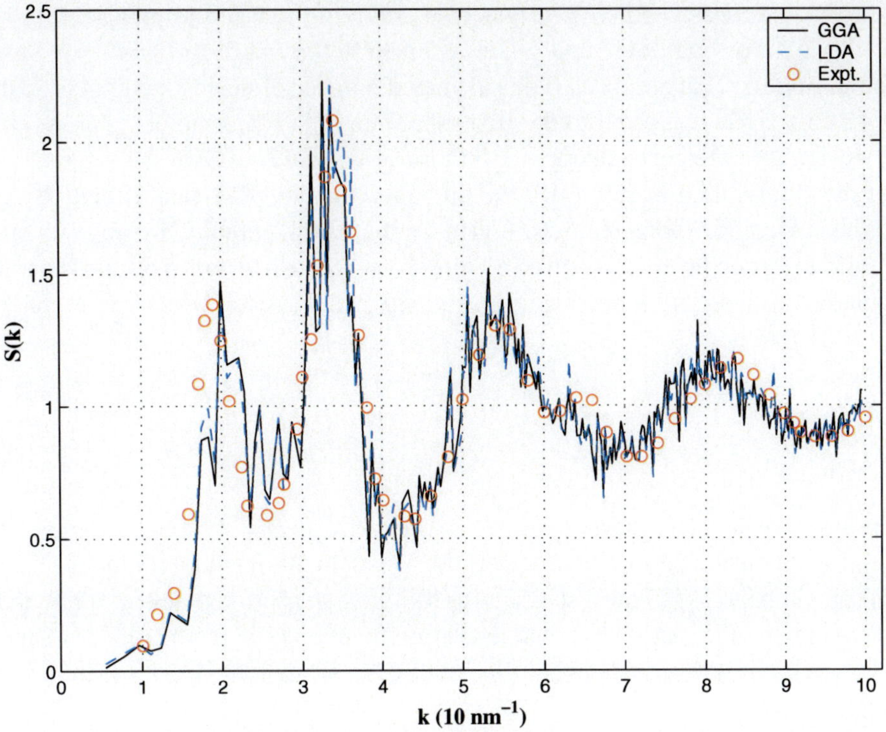


Figure 2: Full curve: Calculated S(k) for $a$-Ge at $T = 300K$, as obtained using the GGA. Structure is prepared as described in the text. Open circles: measured S(k) for $a$-Ge at $T = 300K$, as given in Ref. [52].

We have also calculated S(k) for a model of *amorphous* Ge ($a$-Ge) at $300K$. We prepared our sample of $a$-Ge as described in the previous section. As for $\ell$-Ge, we average the calculated S(k) over different **k** vectors of the same length, as for $\ell$-Ge. In Fig. 2, we show the calculated $S(k)$ for $a$-Ge at T = 300, again using both the GGA and the LDA. The sample is prepared and the averages obtained as described in Section II. In contrast to $\ell$-Ge, but consistent with previous simulations[10, 26], the principal peak in S(k) is strikingly split. The calculations are in excellent agreement with experiments carried out on as-quenched $a$-Ge at $T = 300$ K[52]; in particular, the split principal peak seen in experiment is accurately reproduced by the simulations.

We have also calculated a number of other quantities for both $\ell$-Ge and $a$-Ge, including pair distribution function $g(r)$, and the electronic density of states $n(E)$. For $\ell$-Ge, we calculated $n(E)$ using the Monkhorst-Pack mesh with gamma point shifting (one of the meshes recommended in the VASP package). The resulting $n(E)$ is generally similar to that found in previous calculations[10, 12], provided that an average is taken over at least 5-10 liquid state configurations. Our $n(E)$ for $a$-Ge [calculated using a shorter averaging time than that used below for $S(k,\omega)$] is also similar to that found previously[10]. Our calculated $g(r)$'s for both $\ell$-Ge and $a$-Ge, as given by the VASP program, are similar to those found in Refs. [10] and [12]. The calculated number of nearest neighbors in the first shell is 4.18 for $a$-Ge measured to the first minimum after the principal peak in g(r). For $\ell$-Ge, if we count as "nearest neighbors" all those atoms within 3.4Åof the central atom (the larger of the cutoffs used in Ref. [10]) we find approximately 7.2 nearest neighbors, quite close to the value of 6.9 obtained in Ref. [10] for that cutoff. Finally, we have recalculated the self-diffusion coefficient $D(T)$ for $\ell$-Ge at $T = 1250$ K, from the time derivative of the calculated mean-square ionic displacement; we obtain a result very close to that of Ref. [12].

## 4.2 $S(k,\omega)$ for $\ell$-Ge and $a$-Ge

### 4.2.1 $\ell$-Ge

Fig. 3 shows the calculated ratio $S(k,\omega)/S(k)$ for $\ell$-Ge at $T = 1250$ K, as obtained using the averaging procedure described in Section II. We include a resolution function [eqs. (22) and (23)] of width $\hbar\omega = 2.5$ meV, the same as the quoted experimental width[31]. In Fig. 4, we show the same ratio, but without the resolution function (i. e., with $\omega_0 = 0$). Obviously, there is much more statistical noise in this latter case, though the overall features can still be distinguished.

To interpret these results, we first compare the calculated $S(k,\omega)$ in $\ell$-Ge with hydrodynamic predictions, which should be appropriate at small $k$ and $\omega$. The This prediction takes the form (see, for example, Ref. [53]):

$$\begin{aligned} 2\pi\frac{S(k,\omega)}{S(k)} &= \frac{\gamma-1}{\gamma}\left(\frac{2D_Tk^2}{\omega^2+(D_Tk^2)^2}\right)+ \\ &+ \frac{1}{\gamma}\left(\frac{\Gamma k^2}{(\omega+c_sk)^2+(\Gamma k^2)^2}+\frac{\Gamma k^2}{(\omega-c_sk)^2+(\Gamma k^2)^2}\right). \end{aligned} \tag{24}$$

Here $\gamma = c_P/c_V$ is the ratio of specific heats and constant pressure and constant volume, $D_T$ is the thermal diffusivity, $c_s$ is the adiabatic sound velocity, and $\Gamma$ is the sound attenuation constant. $D_T$ and $\Gamma$ can in turn be expressed in terms of other quantities. For example, $D_T = \kappa_T/(\rho c_P)$, where $\kappa_T$ is the thermal conductivity and $\rho$ is the atomic number density. Similarly, $\Gamma = \frac{1}{2}\left[a(\gamma-1)/\gamma + b\right]$, where $a = \kappa_T/(\rho c_V)$ and $b$ is the kinematic longitudinal viscosity (see, for example, Ref. [53], pp. 264-66).

Eq. (24) indicates that $S(k,\omega)$ in the hydrodynamic regime should have two propagating peaks centered at $\omega = \pm c_s k$, and a diffusive peak centered at $\omega = 0$ and of width determined by $D_T$. The calculated $S(k,\omega)/S(k)$ for the three smallest values of $k$ in Fig. 3, does show the propagating peaks. We estimate peak values of $\hbar\omega \sim 10$ meV for $k = 5.60$ nm$^{-1}$, $\hbar\omega \sim 11$ meV for $k = 7.92$ nm$^{-1}$, and (somewhat less clearly) $\hbar\omega \sim 13$ meV for $k = 9.70$ nm$^{-1}$. The value of $c_s$ estimated from the lowest $k$ value is $c_s \sim 2.7\times 10^5$ cm/sec. (The largest of these three $k$ values may already be outside the hydrodynamic, linear-dispersion regime.)

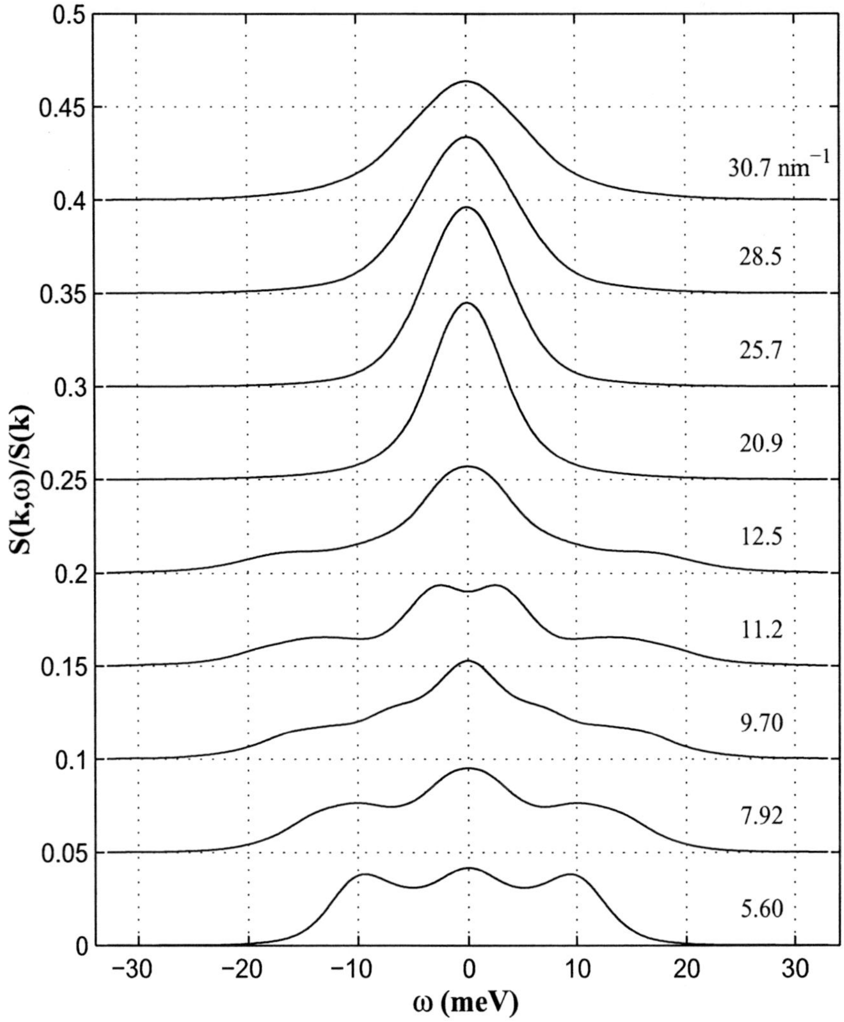


Figure 3: Calculated ratio of dynamic structure factor $S(k, \omega)$ to static structure factor $S(k)$ for $\ell$-Ge at $T = 1250K$ for several values of $k$, plotted as a function of $\omega$, calculated using *ab initio* molecular dynamics with a MD time step of 10 fs. For clarity, each curve has been vertically displaced by 0.05 units from the curve below. In each case, the plotted curve is obtained by averaging both the calculated $S(\mathbf{k}, \omega)$ and the calculated $S(\mathbf{k})$ over all values of $\mathbf{k}$ of the same length. We also incorporate a Gaussian resolution function of half-width $\hbar\omega_0 = 2.5$ meV, as in eqs. (22) and (23). This value of $\omega_0$ is chosen to equal the quoted experimental resolution[31].

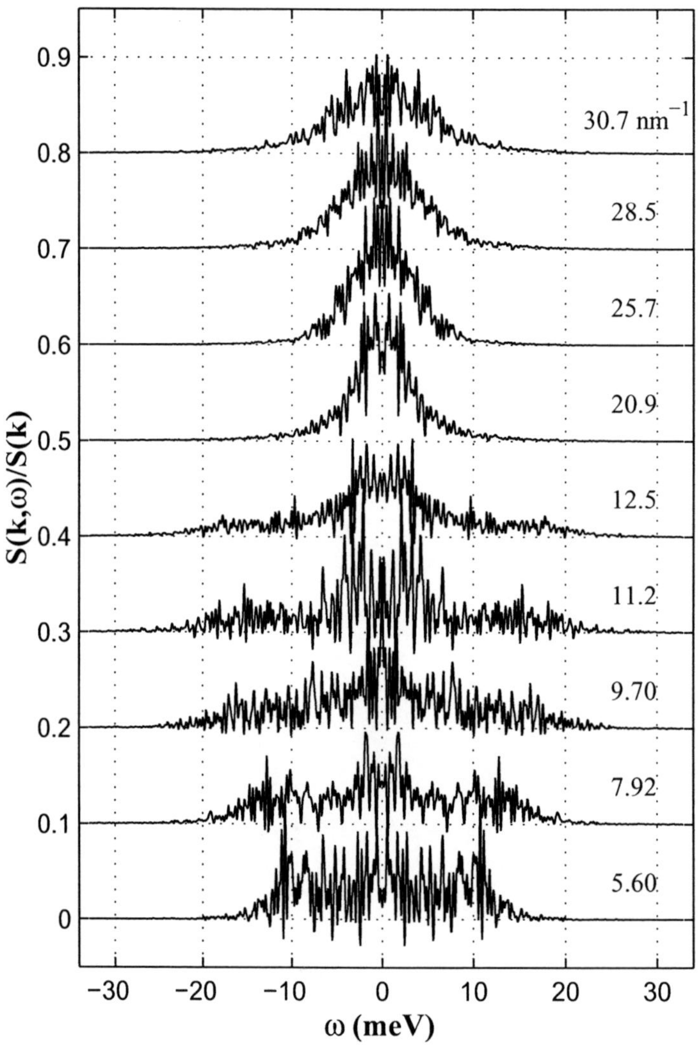


Figure 4: Same as in Fig. 3, but without the resolution function.

These predictions agree reasonably well with the measured $S(k,\omega)$ obtained by Hosokawa *et al*[31], using inelastic X-ray scattering. For example, the measured sound-wave peaks for $k = \pm 6$ nm$^{-1}$ occur near 10 meV, while those $k = \pm 12$ nm$^{-1}$ occur at $\hbar\omega = 17.2$ meV, Furthermore, the integrated relative strength of our calculated sound-wave peaks, compared to that of the central diffusion peak, decreases between $k = 7.92$ nm$^{-1}$ and 12.5 nm$^{-1}$, consistent with both eq. (24) and the change in experimental behavior[31] between $k = 6$ nm$^{-1}$ and 12 nm$^{-1}$.

Because $S(k,\omega)$ in Fig. 3 already includes a significant Gaussian smoothing function, a quantitatively accurate half-width for the central peak, and hence a reliable predicted value for $D_T$, cannot be extracted. A rough estimate can be made as follows. For the smallest k value of 5.6 nm$^{-1}$, the full width of the central peak at half-maximum is around 7.5 meV. If the only broadening were due to this Gaussian smoothing, the full width would be around $2\hbar\omega_0 = 5.5$ meV. Thus, a rough estimate of the intrinsic full width is $\approx \sqrt{7.5^2 - 5.5^2} = 5$ meV $\approx 2\hbar D_T k^2$. This estimate seems reasonable from the raw data for $S(k,\omega)$ shown in Fig. 4. Using this estimate, one obtains $D_T \approx 1.3 \times 10^{-3}$ cm$^2$/sec.

The hydrodynamic expression for $S(k,\omega)/S(k)$ was originally obtained without consideration of the electronic degrees of freedom. Since $\ell$-Ge is a reasonably good metal, one might ask if the various coefficients appearing eq. (24) should be the full coefficients, or just the ionic contribution to those coefficients. For example, should the value of $D_T$ which determines the central peak width be obtained from the full $c_P$, $c_V$, and $\kappa_T$, or from only the ionic contributions to these quantities? For $\ell$-Ge, the question is most relevant for $\kappa_T$, since the dominant contribution to $c_P$ and $c_V$ should be the ionic parts, even in a liquid metal[4]. However, the principal contribution to $\kappa_T$ is expected to be the electronic contribution.

We have made an order-of-magnitude estimate of $D_T$ using the experimental liquid number density and the value $C_P = (5/3)k_B$ per ion, and obtaining the electronic contribution to $\kappa_T$ from the Wiedemann-Franz law[54] together with previously calculated estimates of the electronic contribution[12]. This procedure yields $D_T \sim 0.1$ cm$^2$/sec, about two orders of magnitude greater than that extracted from Fig. 3, and well outside the possible errors in that estimate. We conclude that the $D_T$ which should be used in eq. (24) for $\ell$-Ge (and by inference other liquid metals) is the ionic contribution only.

In support of this interpretation, we consider what one expects for $S(k,\omega)$ in a simple metal such as Na. In such a metal, ionic motions are quite accurately determined by effective pairwise screened ion-ion interactions[4]. Since the ionic motion is determined by such an interaction, the $S(k,\omega)$ resulting from that motion should not involve the contribution of the electron gas to the thermal conductivity. Although $\ell$-Ge is not a simple metal, it seems plausible that its $S(k,\omega)$ should be governed by similar effects, at least in the hydrodynamic regime. This plausibility argument is supported by our numerical results.

For $k$ beyond around 12 nm$^{-1}$, the hydrodynamic model should start to break down, since the dimensionless parameter $\omega\tau$ (where $\tau$ is the Maxwell viscoelastic relaxation time) becomes comparable to unity. At these larger $k$'s, both our calculated and the measured[31] curves of $S(k,\omega)/S(k)$ continue to exhibit similarities. Most notable is the existence of a single, rather narrow peak for $k$ near the principal peak of $S(k)$, followed by a reduction in height and broadening of this central peak as $k$ is further increased. This narrowing was first predicted by de Gennes[55]. In our calculations, it shows up in the plot for $k = 20.9$ nm$^{-1}$, for which the half width of $S(k,\omega)/S(k)$ is quite narrow, while at $k = 28.5$ and $30.7$ nm$^{-1}$, the corresponding plots are somewhat broader and lower. By comparison, the measured central peak in $S(k,\omega)/S(k)$ is narrow at $k = 20$ nm$^{-1}$ and especially at $k = 24$ nm$^{-1}$, while it is broader and lower at $k = 28$ nm$^{-1}$[31].

The likely physics behind the de Gennes narrowing is straightforward. The half-width of $S(k,\omega)$ is inversely proportional to the lifetime of a density fluctuation of wave number $k$. If that $k$ coincides with the principal peak in the structure factor, a density fluctuation will be in phase with the natural wavelength of the liquid structure, and should decay slowly, in comparison to density fluctuations at other wavelengths. This is indeed the behavior observed both in our simulations and in experiment.

In further support of this picture, we may attempt to describe these fluctuations by a very oversimplified Langevin model. We suppose that the Fourier component $\rho(\mathbf{k},t)$ [eq. (18)] is governed by a Langevin equation

$$\dot{\rho}(\mathbf{k},t) = -\xi\rho(\mathbf{k},t) + \eta(t). \tag{25}$$

Here the dot is a time derivative, $\xi$ is a constant, and $\eta(t)$ is a random time-dependent "force" which has ensemble average $\langle\eta\rangle = 0$ and correlation function $\langle\eta(t)\eta^*(t')\rangle = A\delta(t-t')$. Eq. (25) can be solved by standard methods (see, e. g., Ref. [53] for a related example), with the result (for sufficiently large t)

$$\langle\rho(\mathbf{k},t)\rho^*(\mathbf{k},t+\tau)\rangle = \frac{\pi A}{\xi}\exp(-|\tau|\xi). \tag{26}$$

According to eq. (17), $S(k,\omega)$ is, to within a constant factor, the frequency Fourier transform of this expression, i. e.

$$S(k,\omega) \propto \int_{-\infty}^{\infty} (\pi A/\xi)\exp(i\omega\tau)\exp(-|\tau|\xi)d\tau, \tag{27}$$

or, on carrying out the integral,

$$S(k,\omega) \propto \frac{\pi A}{\omega^2+\xi^2}. \tag{28}$$

This is a Gaussian function centered at $\omega = 0$ and of half-width $\xi$. On the other hand, the static structure factor

$$S(k) \propto Lim_{\tau\to 0}\langle\rho(\mathbf{k},t)\rho^*(\mathbf{k},t+\tau)\rangle = \frac{\pi A}{\xi}. \tag{29}$$

Thus, if the constant $A$ is independent of $\mathbf{k}$, the half-width $\xi$ of the function $S(k,\omega)$ at wave number $\mathbf{k}$ is inversely proportional to the static structure $S(k)$. This prediction is consistent with the "de Gennes narrowing" seen in our simulations and in experiment[31].

To summarize, there is overall a striking similarity in the shapes of the experimental and calculated curves for $S(k,\omega)/S(k)$ both in the hydrodynamic regime and at larger values of $k$.

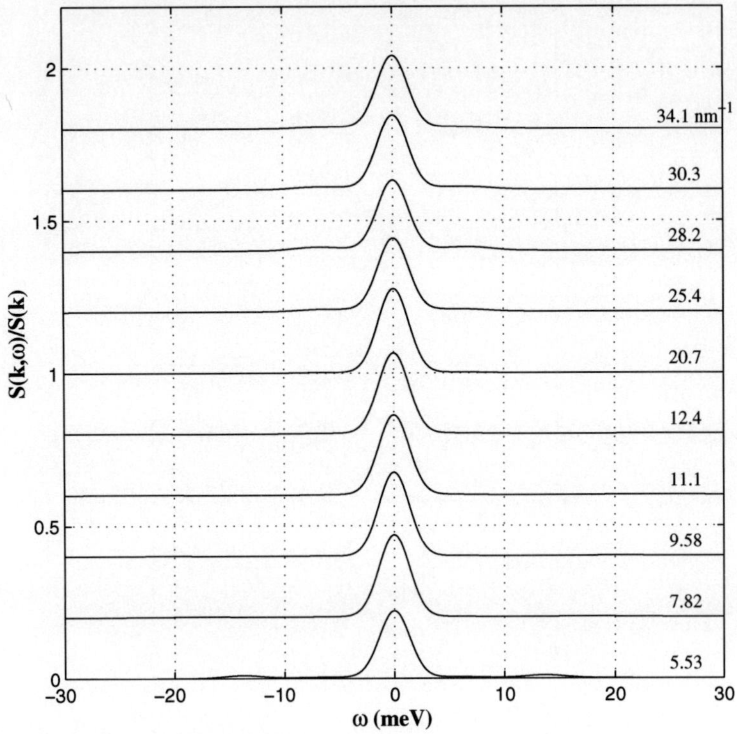


Figure 5: Calculated $S(k,\omega)/S(k)$ for $a$-Ge at $T = 300$ K at $k \leq 35$ nn$^{-1}$, plotted as a function of $\omega$ Again, each curve has been vertically displaced by appropriate amounts from the one below it, as evident from the Figure, and both $S(k,\omega)$ and $S(k)$ have been plotted after an average over all **k**'s of the same length. We also incorporate a Gaussian resolution function of half-width $\hbar\omega_0 = 2$ meV. This value is chosen to give the best results for $n_{vib}(\omega)$ as measured by Ref. [56]. The time step here is 10 fs.

### 4.2.2 $a$-Ge

We have also calculated the dynamic structure for our sample of $a$-Ge at $T = 300K$. The results for the ratio $S(k,\omega)/S(k)$ are shown in Figs. 5 and 6 for a range of $k$ values, and, over a broader range of $\omega$, in Fig. 7. Once again, both $S(k,\omega)$ and $S(k)$ are averaged over different values of **k** of the same length, as described above. We have incorporated a resolution function of width $\hbar\omega_0 = 2$ meV into $S(k,\omega)$. This width is a rough estimate for the experimental resolution function in the measurements of Maley *et al*[56]; we assume it to be smaller than the liquid case because the measured width of the central peak in $S(k,\omega)/S(k)$ for $a$-Ge is quite small.

Ideally, our calculated $S(k,\omega)$ should be compared to the measured one. However, the published measured quantity is not $S(k,\omega)$ but is, instead, based on a modified dynamical

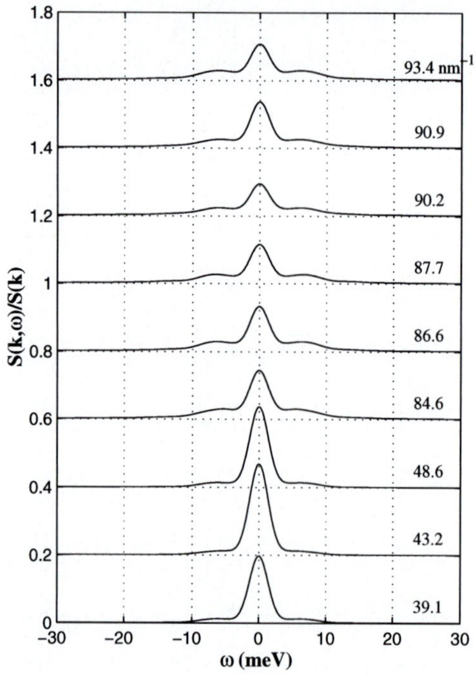


Figure 6: Same as Fig. 5, but at $k \geq 35$ nm$^{-1}$.

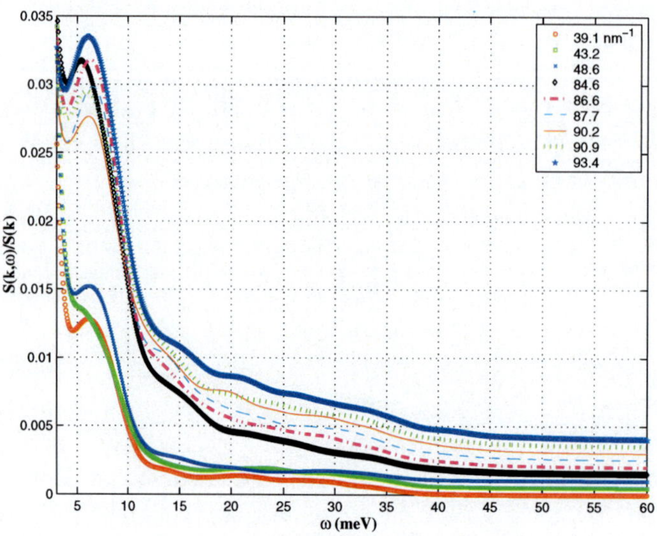


Figure 7: Calculated $S(k,\omega)/S(k)$ as in Figs. 5 and 6 but including higher frequencies $\omega$. At $\hbar\omega = 60$ meV, the curves are arranged vertically in order of increasing frequency. Each curve is vertically displaced by 0.0005 units from the one below it.

structure factor, denoted $G(k,\omega)$, and related to $S(k,\omega)$ by[56]

$$G(k,\omega) = \left(\frac{C}{k^2}\right)\left(\frac{\hbar\omega}{n(\omega,T)+1}\right)S(k,\omega). \tag{30}$$

Here $C$ is a k- and $\omega$-independent constant, and $n(\omega,T) = 1/[e^{\hbar\omega/k_BT} - 1]$ is the phonon occupation number for phonons of energy $E = \hbar\omega$ at temperature $T$. The quantity plotted by Maley *et al*[56] is an average of $G(k,\omega)$ over a range of k values from 40 to 70 nm$^{-1}$. These workers assume that this average is proportional to the vibrational density of states $n_{vib}(\omega)$. The measured $n_{vib}(\omega)$ as obtained in this way[56] is shown in Fig. 8 for two different amorphous structures, corresponding to two different methods of preparation and having differing degrees of disorder.

In order to compare our calculated $S(k,\omega)$ to experiment, we use eq. (30) to infer $G(k,\omega)$, then average over a suitable range of k. However, in using eq. (30), we use the classical form of the occupation factor, $n(\hbar\omega) + 1 \approx k_BT/\hbar\omega$, This choice is justified because we have calculated $S(k,\omega)$ using classical equations of motion for the ions. We thus obtain for the calculated vibrational density of states

$$n_{vib}(\omega) \approx \left(\frac{C\hbar^2}{k_BT}\right)\left(\frac{\omega^2}{k^2}\right)S(k,\omega). \tag{31}$$

In Fig. 8 we show two such calculated plots of $n_{vib}(\omega)$, as obtained by averaging eq. (31) over two separate groups of $k$'s, as indicated in the caption[57]. For comparison, we also show $n_{vib}(\omega)$ for $a$-Ge as calculated in Ref. [10] directly from the Fourier transform of the velocity-velocity autocorrelation function.

The calculated plots for $n_{vib}(\omega)$ in Fig. 8 have some distinct structure, which arises from some corresponding high frequency structure in $S(k,\omega)$. The plot of $n_{vib}(\omega)$ for the group of smaller k's has two distinct peaks, near 8 meV and 29 meV, separated by a broad dip with a minimum near 18 meV. The plot corresponding to the group of larger k's has similar structure and width, but the dip is less pronounced. The two experimental plots also have two peaks separated by a clear dip. The two maxima are found around 10 and 35 meV, while the principal dip occurs near 16 meV. In addition, the overall width of the two densities of states is quite similar.

The reasonable agreement between the calculated and measured $n_{vib}(\omega)$ suggests that our *ab initio* calculation of $S(k,\omega)$ for $a$-Ge is reasonably accurate. The noticeable differences probably arise from several factors. First, there are several approximations involved in going from the calculated and measured $S(k,\omega)$'s to the corresponding $n_{vib}(\omega)$'s, and these may be responsible for some of the discrepancies. Secondly, there may actually be differences between the particular amorphous structures studied in the experiments, and the quenched, then relaxed structure considered in the present calculations. (However, the similarities in the *static* structure factors suggest that these differences are not vast.) Finally, our calculations are carried out over relatively short times, using relatively few atoms; thus, finite-size and finite-time effects are likely to produce some additional errors. Considering all these factors, agreement between calculation and experiment is quite reasonable.

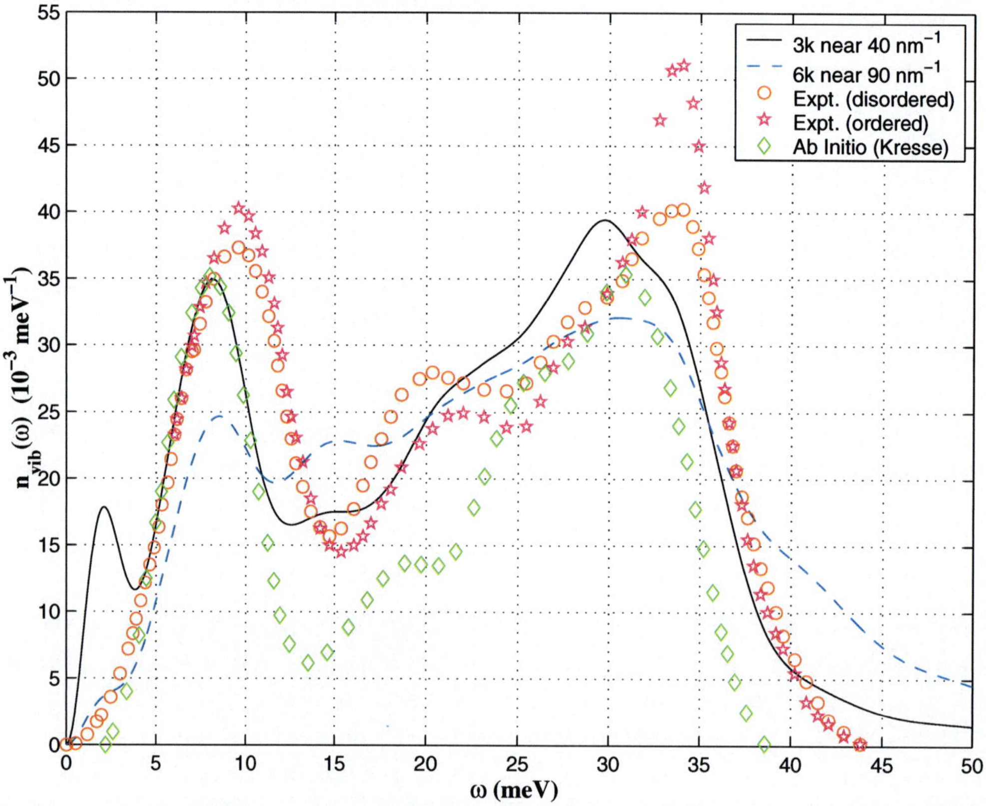


Figure 8: Full curve: calculated vibrational density of states $n_{vib}(\omega)$, in units of $10^{-3}$ states/meV. $n_{vib}(\omega)$ is obtained from the resolution-broadened $S(k,\omega)$ and $S(k)$ of the previous Figure, using the formula $n_{vib}(\omega) = \langle G_{calc}(k,\omega)\rangle$, where $G_{calc}(k,\omega)$ is given by eq. (16), and the averaging is carried out over the three magnitudes of $k$ near 40 nm$^{-1}$ for which we have computed $S(k,\omega)$. Dashed curve: same as full curve, but calculated by averaging over the six magnitudes of $k$ near 90 nm$^{-1}$ for which we have computed $S(k,\omega)$. The open circles and open stars denote the measured $n_{vib}(\omega)$, as reported in Ref. [56] for two forms of $a$-Ge. Finally, the open diamonds denote $n_{vib}(\omega)$ as calculated in Ref. [10] from the ionic velocity-velocity autocorrelation function (dot-dashed curves). In all plots except that of Ref. [10], $n_{vib}(\omega)$ is normalized so that $\int_0^{\omega_{max}} n(\omega)d(\hbar\omega) = 1$. $\omega_{max}$ is the frequency at which $n_{vib}(\omega) \rightarrow 0$, and is estimated from this Figure by extrapolating the right hand parts of the solid and dashed curves linearly to zero.

Previous *ab initio* calculations for $a$-Ge[10] have also obtained a vibrational density of states, but this is computed directly from the ionic velocity-velocity autocorrelation function rather than from the procedure described here. The calculations in Ref. [10] do not require computing $S(k, \omega)$. In the present work, by contrast, we start from our calculated $S(k, \omega)$, and we work backwards to get $n_{vib}(\omega)$. In principle, our $S(k, \omega)$ includes all anharmonic effects on the vibrational spectrum of $a$-Ge, though in extracting $n_{vib}(\omega)$ we assume that the lattice vibrates harmonically about the metastable atomic positions. In Fig. 8, we also show the results of Ref. [10] for $n_{vib}(\omega)$ as obtained from this correlation function. They are quite similar to those obtained in the present work, but have a somewhat deeper minimum between the two principal peaks.

The quantity $n_{vib}(\omega)$ could, of course, also be calculated directly from the force constant matrix, obtained by assuming that the quenched configuration is a local energy minimum and calculating the potential energy for small positional deviations from that minimum using *ab initio* molecular dynamics. This procedure has been followed for $a$-GeSe$_2$, for example, by Cappelletti *et al*[58]. These workers have then obtained $S(q, \omega)$ versus $q$ from their $n_{vib}(\omega)$ at selected values of $\omega$, within a one-phonon approximation. However, as noted above, the present work produces the full $S(k, \omega)$ and thus has, in principle, more information than $n_{vib}(\omega)$.

# 5 Discussion and Conclusions

The results reviewed here show that *ab initio* molecular dynamics can be used to calculate the dynamic structure factor $S(k, \omega)$ for both liquid and amorphous semiconductors. Although the accuracy of the calculated $S(k, \omega)$ is lower than that attained for static quantities, such as $S(k)$, nonetheless it is sufficient for comparison to most experimental features. This is true even though our calculations are limited to 64-atom samples and fewer than 20 ps of elapsed real time.

We have presented evidence that the calculated $S(k, \omega)/S(k)$ in $\ell$-Ge agrees qualitatively with measured by inelastic X-ray scattering[31], and that the one calculated for $a$-Ge leads to a vibrational density of states qualitatively similar to the quoted experimental one[56]. Since such calculations are thus shown to be feasible, the work reviewed here should spur further numerical studies, with longer runs on larger samples, to obtain even more detailed information. Furthermore, we can use these dynamical simulations to probe the underlying processes at the atomic scale which give rise to specific features in the measured and calculated $S(k, \omega)$.

**Acknowledgements**

This work has been supported by NSF grants DMR01-04987 (D.S.), DMR04-12295 (D.S.) and CHE01-11104 (J.D.C.). Calculations were carried out using the Beowulf Cluster at the Ohio Supercomputer Center, with the help of a grant of time. Jeng-Da Chai acknowledges full support from the UMCP Graduate School Fellowship, the IPST Alexander Family Fellowship, and the CHPH Block Grant Supplemental Fellowship.

## References


[1] V. M. Glazov, S. N. Chizhevskaya, and N. N. Glagoleva, *Liquid Semiconductors* (Plenum, New York, 1969).

[2] W. B. Yu, Z. Q. Wang, and D. Stroud, *Phys. Rev.* **B54**, 13946 (1996).

[3] F. H. Stillinger and Weber, *Phys. Rev.* **B31**, 5262 (1985).

[4] For a review, see, e. g., N. W. Ashcroft and D. Stroud, *Solid State Physics* **33**, pp. 1ff (1978).

[5] N. W. Ashcroft, *Il Nuovo Cimento* **12D**, 597 (1990).

[6] P. Hohenberg and W. Kohn, *Phys. Rev.* **136**, B864 (1964).

[7] W. Kohn and L. J. Sham, *Phys. Rev.* **140**, A1133 (1965).

[8] N. D. Mermin, *Phys. Rev.* **137**, A1441 (1965).

[9] R. Car and M. Parrinello, *Phys. Rev. Lett.* **35**, 2471 (1985).

[10] G. Kresse and J. Hafner, *Phys. Rev.* **B49**, 14251 (1994).

[11] N. Takeuchi and I. L. Garzón, *Phys. Rev.* **B50**, 8342 (1995).

[12] R. V. Kulkarni, W. G. Aulbur, and D. Stroud, *Phys. Rev.* **B55**, 6896 (1997).

[13] R. V. Kulkarni and D. Stroud, *Phys. Rev.* **B57**, 10476 (1998).

[14] Q. Zhang, G. Chiarotti, A. Selloni, R. Car, and M. Parrinello, *Phys. Rev.* **B42**, 5071.

[15] L. Lewis, A. De Vita, and R. Car, *Phys. Rev.* **B57**, 1594 (1998).

[16] R. V. Kulkarni and D. Stroud, *Phys. Rev.* **B62**, 4991 (2000).

[17] V. Godlevsky, J. Derby, and J. Chelikowsky, *Phys. Rev. Lett.* **81**, 4959 (1998); V. Godlevsky, M. Jain, J. Derby, and J. Chelikowsky, *Phys. Rev.* **B60**, 8640 (1999);

[18] M. Jain, V. V. Godlevsky, J. J. Derby, and J. R. Chelikowsky, *Phys. Rev.* **B65**, 035212 (2001).

[19] R. Car and M. Parrinello, *Phys. Rev. Lett.* **63**, 204 (1988)

[20] I Stich, R. Car, and M. Parrinello, *Phys. Rev. Lett.* **63**, 2240 (1989); *Phys. Rev.* **B44**, 4261 (1991); *Phys. Rev.* **B44**, 11092 (1991).

[21] I. Lee and K. J. Chang, *Phys. Rev.* **B50**, 18083 (1994).

[22] N. C. Cooper, C. M. Goringe, and D. R. McKenzie, *Comput. Mater. Sci.* **17**, 1 (2000)

[23] See, e. g., J. P. Perdew, in *Electronic Structure of Solids*, edited by P. Ziesche and H. Eschrig (Akademic Verlag, Berlin, 1991).

[24] O. Sankey and D. J. Niklewsky, *Phys. Rev.* **B40**, 3979 (1989)

[25] D. A. Drabold, P. A. Fedders, O. F. Sankey, and J. D. Dow, *Phys. Rev.* **B42**, 5135 (1990)

[26] F. Alvarez, C. C. Díaz, A. A. Valladares, and R. M. Valladares, *Phys. Rev.* **B65**, 113108 (2002).

[27] C. Z. Wang, C. T. Chan, and K. M. Ho, *Phys. Rev.* **B45**, 12227 (1992); I. Kwon, R. Biswas, C. Z. Wang, K. M. Ho, and C. M. Soukoulis, *Phys. Rev.* **B49**, 7242 (1994).

[28] G. Servalli and L. Colombo, *Europhys. Lett.* **22**, 107 (1993).

[29] C. Molteni, L. Colombo, and Miglio, *J. Phys.: Cond. Matt.* **6**, 5255 (1994).

[30] J.-D. Chai, D. Stroud, J. Hafner, and G. Kresse, *Phys. Rev.* **B67**, 104205 (2003).

[31] S. Hosokawa, Y. Kawakita, W.-C. Pilgrim, and H. Sinn, *Phys. Rev.* **B63**, 134205 (2001).

[32] R. G. Parr and W. Yang, Density-Functional Theory of Atoms and Molecules, (Oxford University Press, 1989).

[33] R. M. Dreizler and E. K. U. Gross, Density Functional Theory: An Approach to the Quantum Many Body Problem, (Springer-Verlag, Berlin, 1990).

[34] R. A. King and N. C. Handy, *Phys. Chem. Chem. Phys.* **2**, 5049 (2000).

[35] D. G. Kanhere, P. V. Panat, A. K. Rajagopal, and J. Callaway, *Phys. Rev.* **A33**, 490 (1986).

[36] L. H. Thomas, *Proc. Cambridge Phil. Soc.* **23**, 542 (1927).

[37] E. Fermi, *Z. Phys.* **48**, 73 (1928).

[38] See e.g., Y. A. Wang and E. A. Carter, in *Theoretical Methods in Condensed Phase Chemistry*, edited by S.D. Schwartz, *Progress in Theoretical Chemistry and Physics*, (Kluwer, Boston, 2000) p. 117, and references therein.

[39] L.-W. Wang and M. P. Teter, *Phys. Rev.* **B45**, 13196 (1992).

[40] Y. A. Wang, N. Govind, and E. A. Carter, *Phys. Rev.* **B58**, 13465 (1998).

[41] Y. A. Wang, N. Govind, and E. A. Carter, *Phys. Rev.* **B60**, 16350 (1999).

[42] J.-D. Chai and J. D. Weeks, *J. Phys. Chem.* **B108**, 6870 (2004).

[43] J.-D. Chai and J. D. Weeks, (unpublished).

[44] M. Methfessel and A. T. Paxton, *Phys. Rev.* **B40**, 3616 (1989).

[45] M. Weinert and J. W. Davenport, *Phys Rev.* **B45**, 13709 (1992).

[46] R. M. Wentzcovitch, J. L. Martins, and P. B. Allen, *Phys. Rev.* **B45**, 11372 (1992).

[47] G. Kresse and J. Hafner, *Phys. Rev.* **B47**, 558 (1993); G. Kresse, Thesis, Technische Universität Wien (1993); G. Kresse and J. Furthmüller, *Comput. Mat. Sci.* **6**, pp. 15-50 (1996); G. Kresse and J. Furthmüller, *Phys. Rev.* **B54**, 11169 (1996).

[48] D. Vanderbilt, *Phys. Rev.* **B32**, 8412 (1985).

[49] H. J. Monkhorst and J. D. Pack, *Phys. Rev.* **B13**, 5188 (1976).

[50] Our method is very similar to that used in Ref. [10] to treat $\ell$-Ge and $a$-Ge, but with the following differences: (i) we use both the GGA and the LDA, while Ref. [10] uses only the LDA; (ii) we include slightly more bands (161 as compared to 138); and (iii) most of our calculations are carried out using a 10 fs time step, rather than the 3 fs step used in Ref. [10]. Our $a$-Ge structure may also differ slightly from that of Ref. [10], since we use a somewhat faster quench rate ($\sim 3 \times 10^{14}$ K/sec rather than $1.7 \times 10^{14}$ K/sec). Both codes use the conjugate gradient method to find energy minima and the same Vanderbilt ultrasoft pseudopotentials.

[51] Y. Waseda, The Structure of Noncrystalline Materials: Liquids and Amorphous Solids (McGraw-Hill, 1980).

[52] G. Etherington, A. C. Wright, J. T. Wenzel, J. T. Dore, J. H. Clarke, and R. N. Sinclair, *J. Non-Cryst. Solids* **48**, 265 (1982).

[53] See, for example, J.-P. Hansen and I. R. McDonald, *Theory of Simple Liquids*, 2nd edition (Academic Press, London, 1986), pp. 222-228.

[54] See, for example, N. W. Ashcroft and N. D. Mermin, *Solid State Physics* (Harcourt-Brace, Fort Worth, 1976), ch. 2.

[55] P. G. de Gennes, *Physics* **25**, 825 (1959); *Physics* **3**, 37 (1967).

[56] N. Maley, J. S. Lannin, and D. L. Price, *Phys. Rev. Lett.* **56**, 1720 (1986).

[57] Two recent groups [V. Martín-Mayor, M. Mézard, G. Parisi, and P. Verrocchio, *J. Chem. Phys.* **114**, 8068 (2001), and T. S. Grigera, V. Martín-Mayor, G. Parisi, and P. Verrocchio, *Phys. Rev. Lett.* **87**, 085502 (2001)] have derived approximate expressions

connecting $S(k, \omega)$ to $n_{vib}(\omega)$ in an amorphous solid, which closely resemble eq. (31). In fact, their expressions can be shown to become equivalent to eq. (31) at sufficiently large k.

[58] R. L. Cappelletti, M. Cobb, D. A. Drabold, and W. A. Kamitakahara, *Phys. Rev.* **B52**, 9133 (1995).

In: Progress in Chemical Physics Research
Editor: A. N. Linke, pp. 107-114

ISBN: 1-59454-451-4


*Chapter 5*

# NEGATIVE THERMAL EXPANSION OF $ZrW_2O_8$ AND $HfW_2O_8$: CONTRADICTION TO A THERMODYNAMIC RELATION

***I. A. Stepanov***[1*]
Institute of Physical Chemistry, University of Freiburg, Albertstrasse 23a, Freiburg 79104, Germany

## Abstract

It is shown that the dependence of negative thermal expansion coefficient of $ZrW_2O_8$ and $HfW_2O_8$ on the temperature found by many authors contradicts to an important thermodynamic relation $(\partial C_P/\partial P)_T = - TV((\partial \alpha/\partial T)_P + \alpha^2)$.

**Keywords:** negative thermal expansion, heat capacity, $ZrW_2O_8$, $HfW_2O_8$, the 1st law of thermodynamics

In [1-9] dependence of the volume on the temperature and heat capacity $C_P$ on the temperature has been obtained for $ZrW_2O_8$ and $HfW_2O_8$. Both substances have cubic crystal structure. There is a thermodynamic relation using which one can verify the 1st law of thermodynamics [10, 11]:

$$(\partial C_P/\partial P)_T = - TV((\partial \alpha/\partial T)_P+\alpha^2) \quad (1)$$

where $\alpha$ is the thermal expansion coefficient. One can show that the dependence $\alpha(T, P)$ found in [1-6] contradicts to the relation (1) for negative $\alpha$.

In [1-6] for both substances, $(\partial \alpha/\partial T)_P < 0$ in big temperature intervals because $V \approx V_0 - bT$, b = const. If $V \approx V_0 - bT$ then $(\partial \alpha/\partial T)_P + \alpha^2 \approx - b^3T/V_0^3$. The derivative $(\partial C_P/\partial T)_P > 0$ in

[1] Permanent address: Juglas 1a-20, Riga, LV-1024, Latvia.
[*] E-mail address: istepanov@email.com

the same temperature intervals. Heat capacity $C_P$ increases with the temperature from $T \approx 300$ K till $T = T_{max}$ ($T_{max} \approx 440$ K for $ZrW_2O_8$), then it decreases (till $T \approx 470$ K for $ZrW_2O_8$), then it again increases (at $T > 470$ K for $ZrW_2O_8$) [2, 7], Figure 1. In [8, 9] $C_P(T)$ has been measured from 0 till 300 K. It increases drastically from T = 0 K till $T \approx 100$ K, Figure 2.

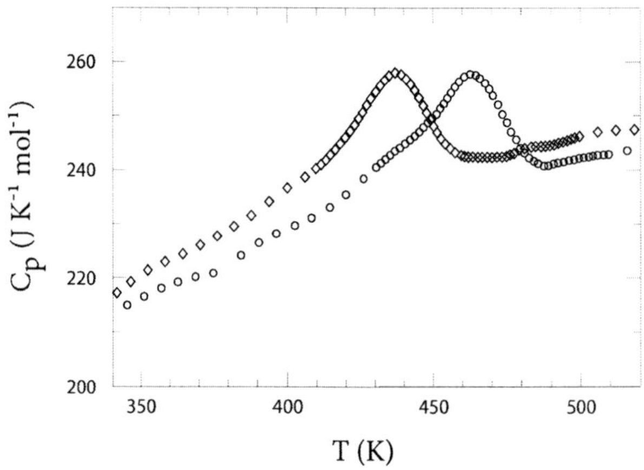


Figure 1. Heat capacities of $HfW_2O_8$ (°) and $ZrW_2O_8$ (◊) as a function of temperature [2].

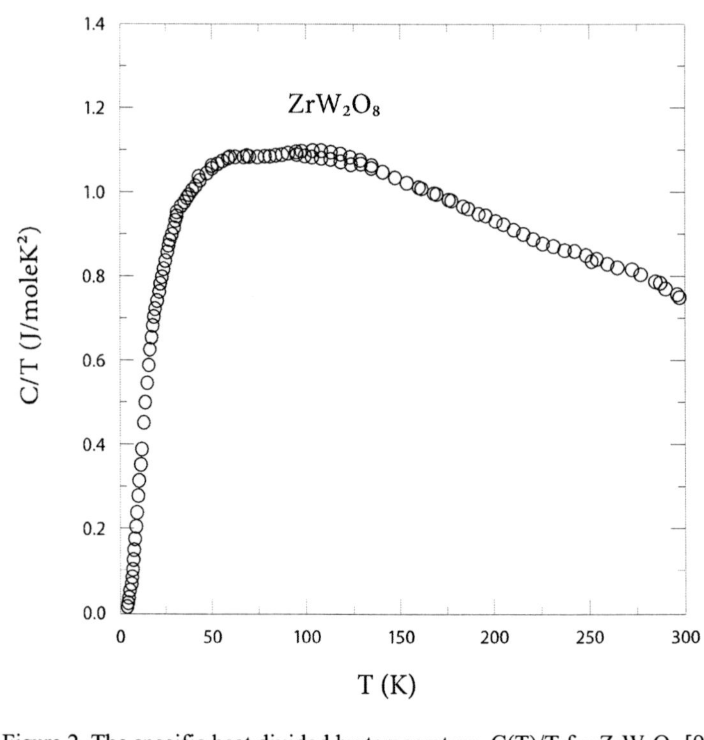


Figure 2. The specific heat divided by temperature, C(T)/T for $ZrW_2O_8$ [9].

In Tables 1-3 and Figures 3 and 4 dependence of the lattice parameter and α of $ZrW_2O_8$ and $HfW_2O_8$ on the temperature is given. The data from Table 1 are from [3], they were provided by Prof. Evans, the data from Table 2 are from [4], they were provided by Prof.

Sleight, the data from Table 3 are from [2], they were provided by Prof. Tsuji, Figure 4 was provided by Prof. Hashimoto. The data in [5] coincide with the data from [1] within the error limits in the interval 300 < T < 420 K. The dependence α(T) is found using numerical differentiation of the data.

From Tables 1 and 2 and Figures 3 and 4 one can deduce that: at 0 < T < 100 K α of $ZrW_2O_8$ steeply decreases. From T ≈ 400 K till $T = T_{min} \approx 440$ K α decreases drastically, then it steeply increases till T ≈ 470 K, then it again decreases. There $|(\partial\alpha/\partial T)_P| >> \alpha^2$.

Table 1. Dependence of the lattice constant of $ZrW_2O_8$ on the temperature [3]

| T, K | a, Å | $\alpha = 1/V(\partial V/\partial T)_P$, *$10^4$, $K^{-1}$ |
|---|---|---|
| 2 | 9.1800 | 0.0395 |
| 20 | 9.1797 | -0.1367 |
| 40 | 9.1785 | -0.2364 |
| 60 | 9.1769 | -0.2903 |
| 80 | 9.1750 | -0.3186 |
| 100 | 9.1731 | -0.2991 |
| 120 | 9.1713 | -0.3001 |
| 140 | 9.1694 | -0.3159 |
| 160 | 9.1675 | -0.3012 |
| 180 | 9.1657 | -0.2955 |
| 200 | 9.1639 | -0.2842 |
| 220 | 9.1622 | -0.2864 |
| 240 | 9.1604 | -0.2894 |
| 260 | 9.1587 | -0.2755 |
| 280 | 9.1570 | -0.2791 |
| 300 | 9.1553 | -0.2789 |
| 320 | 9.1536 | -0.2766 |
| 340 | 9.1519 | -0.2862 |
| 360 | 9.1501 | -0.2995 |
| 380 | 9.1484 | -0.2372 |
| 400 | 9.1470 | -0.2767 |
| 420 | 9.1447 | -0.4762 |
| 440 | 9.1432 | -0.4752 |
| 460 | 9.1416 | -0.6746 |
| 480 | 9.1394 | -0.5675 |
| 500 | 9.1384 | -0.2066 |
| 520 | 9.1372 | -0.1859 |

Table 2. Dependence of the lattice constant of $ZrW_2O_8$ on the temperature [4]

| T, K | a, Å | $\alpha = 1/V(\partial V/\partial T)_P$, $*10^4$, $K^{-1}$ |
|---|---|---|
| 2 | 9.1800 | 0.0577 |
| 22 | 9.1796 | -0.1655 |
| 42 | 9.1783 | -0.2534 |
| 62 | 9.1767 | -0.2770 |
| 82 | 9.1749 | -0.3059 |
| 102 | 9.1730 | -0.3042 |
| 122 | 9.1711 | -0.3071 |
| 142 | 9.1692 | -0.3172 |
| 162 | 9.1673 | -0.3086 |
| 182 | 9.1654 | -0.2893 |
| 202 | 9.1637 | -0.2871 |
| 222 | 9.1619 | -0.2859 |
| 242 | 9.1602 | -0.2784 |
| 262 | 9.1585 | -0.2806 |
| 282 | 9.1568 | -0.2797 |
| 302 | 9.1551 | -0.2764 |
| 322 | 9.1534 | -0.2811 |
| 342 | 9.1517 | -0.2706 |
| 362 | 9.1501 | -0.2986 |
| 382 | 9.1482 | -0.2561 |
| 402 | 9.1468 | -0.2902 |
| 422 | 9.1444 | -0.4673 |
| 442 | 9.1412 | -0.5572 |
| 462 | 9.1382 | -0.3410 |
| 482 | 9.1371 | -0.1370 |
| 502 | 9.1360 | -0.1945 |
| 520 | 9.1349 | -0.2204 |

The data for $HfW_2O_8$ in Table 3 are insufficient to built a smooth curve but a drastic minimum at $T \approx 460$ K is evident. The dependence $C_P(T)$ has a drastic maximum at this point [2], Figure 1.

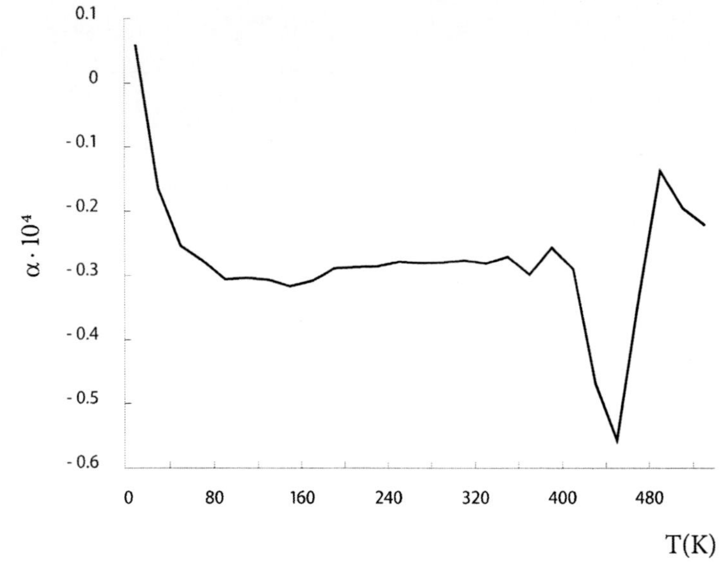


Figure. 3. Dependence of the thermal expansion coefficient of $ZrW_2O_8$ on the temperature (Table 2).

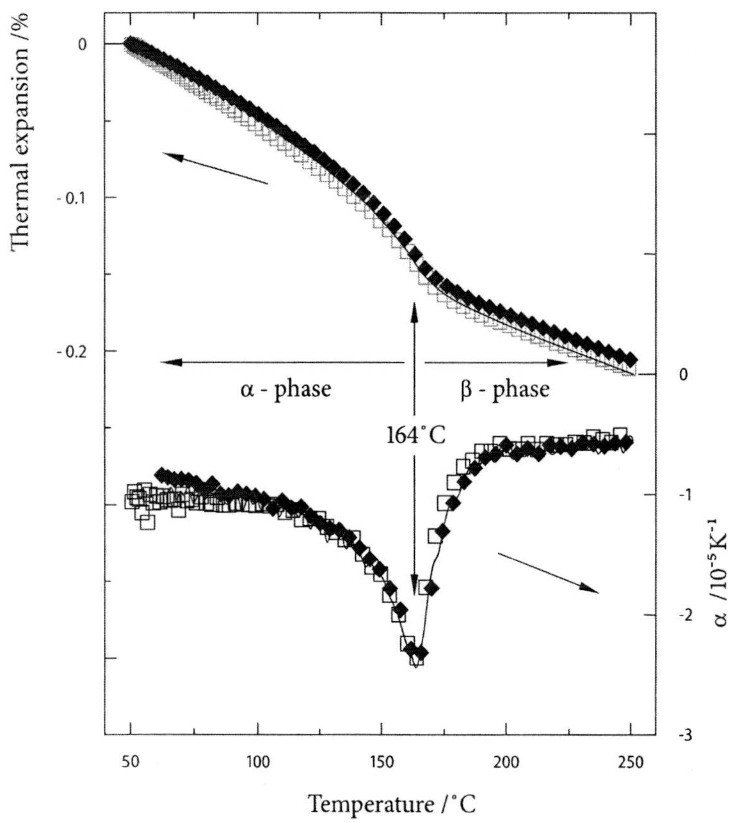


Figure 4. Thermal expansion and thermal expansion coefficient, α, of $ZrW_2O_8$ ceramic specimen measured with dilatometer. A λ-type transition was observed at 164 °C. Stable $ZrW_2O_8$ phase is depicted in the Figure [6].

Table 3. Dependence of the lattice constant of $HfW_2O_8$ on the temperature [2]

| T, K | a, Å | $\alpha = 1/V(\partial V/\partial T)_P$, $*10^4$, $K^{-1}$ |
|---|---|---|
| 91 | 9.1469 | -0.3075 |
| 121 | 9.1441 | -0.3056 |
| 171 | 9.1396 | -0.2867 |
| 220 | 9.1357 | -0.2158 |
| 270 | 9.1325 | -0.2669 |
| 299 | 9.1300 | -0.2390 |
| 319 | 9.1286 | -0.2904 |
| 338 | 9.1266 | -0.3184 |
| 358 | 9.1251 | -0.2096 |
| 377 | 9.1238 | -0.2665 |
| 396 | 9.1221 | -0.2549 |
| 415 | 9.1209 | -0.1970 |
| 435 | 9.1194 | -0.3440 |
| 454 | 9.1168 | -0.5422 |
| 473 | 9.1138 | -0.3853 |
| 501 | 9.1122 | -0.1467 |
| 520 | 9.1111 | -0.2010 |
| 539 | 9.1101 | -0.1514 |
| 558 | 9.1094 | -0.0894 |

However, one can prove that

$$\text{sign}\,(\partial C_P/\partial P)_T = \text{sign}\,dC_P/dP = \text{sign}\,(dC_P/dT)(dT/dP). \quad (2)$$

Heat capacity is the function of pressure and temperature: $C_P = C_P(P, T)$. A special case is $C_P = C_P(T(P))$. The derivative $dC_P/dP$ has a certain sign at $T \approx$ const in a small vicinity of the point T = const. At the point T = const it will have the same sign due to continuity of the function $C_P(P, T)$.

One can prove that $dC_P/dP$ does not change its sign. There is the Rolle's lemma: If f(x) is continious in the interval $a \le x \le b$, and if $f(a) > 0$ and $f(b) < 0$ then there exists at least one value of x (say x = c) such that f(c) = 0, where $a < c < b$. Let us suppose that $dC_P/dP$ changes its sign outside the vicinity of the point T = const. Then at the border of this vicinity, $dC_P/dP = 0$. But $dC_P = (\partial C_P/\partial P)_T dP + (\partial C_P/\partial T)_P dT$. For $ZrW_2O_8$ and $HfW_2O_8$, $(\partial\alpha/\partial T)_P < 0$ in big temperature intervals (Figure 3), therefore, according to (1), $(\partial C_P/\partial P)_T > 0$. In the same temperature intervals, $(\partial C_P/\partial T)_P > 0$ (Figures 1, 2, 4). Thus $dC_P$ can not be equal to 0.

The derivative dT/dP for $ZrW_2O_8$ and $HfW_2O_8$ has the same sign as $(\partial T/\partial P)_V$ thus it is less than 0. Due to continuity of the function T, in a small vicinity of the point $V \approx$ const, dT/dP has the same sign as $(\partial T/\partial P)_V$. Let us suppose that outside this vicinity it is greater than 0. Then from the Rolle's lemma, at the border of this vicinity, dT/dP = 0 or $dP/dT = \infty$. It does not happen in reality.

Analogously one can show that $dC_P/dT$ also does not change its sign which equals the sign of $(\partial C_P/\partial T)_P$. The differential $dC_P$ can not be equal to 0.

Let us consider the temperature interval $0 < T < 100$ K for $ZrW_2O_8$. For $\alpha < 0$, $dT/dP < 0$ and $dC_P/dT > 0$ thus the left part of (1) $(\partial C_P/\partial P)_T < 0$. The right side of (1) has positive sign. It is a contradiction. In the temperature interval $400 < T < 440$ K for $ZrW_2O_8$ $(\partial C_P/\partial P)_T < 0$ but the right side of (1) is positive. In the temperature interval $440 < T < 470$ K $(\partial C_P/\partial P)_T > 0$ but the right side of (1) is negative. At $T > 470$ K, $(\partial C_P/\partial P)_T < 0$ but $-(\partial\alpha/\partial T)_P > 0$.

A possible explanation of this paradox can be found in [11-14]. There it has been supposed that for substances with negative thermal expansion the 1st law of thermodynamics must have the following form: $\delta Q = dU - PdV$. If to derive (1) using this formula, one obtains

$$(\partial C_P/\partial P)_T = TV((\partial\alpha/\partial T)_P + \alpha^2). \tag{3}$$

Notice the change of sign with respect to (1). (In [11] there is a misprint in equation (11). The correct equation is: for $\alpha < 0$, $C_V = C_P k_S/k_T$).

# References


[1] Evans J. S. O., Mary T. A., Vogt T., Subramanian M. A. and Sleight A. W. (1996) Negative Thermal Expansion in $ZrW_2O_8$ and $HfW_2O_8$. *Chem. Mater.* **8**, 2809-2823

[2] Yamamura Y., Nakajima N. and Tsuji T. (2001) Calorimetric and x-ray diffraction studies of $\alpha$ - to - $\beta$ structural phase transitions in $HfW_2O_8$ and $ZrW_2O_8$. *Phys. Rev. B* **64**, 184109

[3] Evans J. S. O., David W. I. F. and Sleight A. W. (1999) Structural Investigation of the Negative - Thermal - Expansion Material $ZrW_2O_8$. *Acta Crystallogr.* **B55**, 333-340

[4] Duan N., Kameswari U. and Sleight A. W. (1999) Further Contraction of $ZrW_2O_8$. *J. Am. Chem. Soc.* **121**, 10432-10433

[5] Hashimoto T., Katsube T. and Morito Y. (2000) Observation of Two Kinds of Phase Transitions of $ZrW_2O_8$ by Power-Compensated Differential Scanning Calorimetry and High-Temperature X-ray Diffraction. *Sol. St. Comm.* **116**, 129-132

[6] Hashimoto T. and Morito Y. (2002) Thermal Analysis of Phase Transition of Negative - Thermal - Expansion Oxide, $ZrW_2O_8$ – Detection of Trace Amount of $H_2O$ and $\lambda$-type Transition. *J. Ceram. Soc. Jpn.*, **110** (9), 823-825

[7] Yamamura Y., Nakajima N. and Tsuji T. (2000) Heat Capacity Anomaly due to the alpha -to - beta Structural Phase Transition in $ZrW_2O_8$. *Sol. St. Comm.* **114**, 453-455

[8] Yamamura Y., Nakajima N., Tsuji T., Iwasa Y., Saito K. and Sorai M. (2002) Heat Capacity and Grüneisen Function of Negative Thermal Expansion Compound $HfW_2O_8$. *Sol. St. Comm.* **121**, 213-217

[9] Ramirez A. P. and Kowach G. R. (1998) Large Low Temperature Specific Heat in the Negative Thermal Expansion Compound $ZrW_2O_8$. *Phys. Rev. Lett.* **80**, 4903-4906

[10] Guemez J., Fiolhais C. and Fiolhais M. (1999) Thermodynamics at Work: The Pressure Derivative of the Specific Heat. *Am. J. Phys.* **67**, 1100-1104

[11] Stepanov I. A. (2000) Thermodynamics of Substances with Negative Thermal Expansion Coefficient. *Computer Modelling & New Technologies* **4**, No 2, 72-74. (available at www.tsi.lv)

[12] Stepanov I. A. (2004) Thermodynamics of Iodine under High Pressure at Low Temperatures. *Physica B* **349**, 1-4, 251-253

[13] Stepanov I. A. (2005) $(\partial S/\partial V)_U$ is negative for substances with negative thermal expansion. - Trends in Chemical Physics Research. Ed. A. N. Linke. Nova Science Publishers, Inc., Hauppauge, NY, 2005. In press.

[14] Stepanov I. A. (2005) The Difference Between the Heats of Chemical Reactions Obtained by the Van't-Hoff Equation and by Calorimetry. - Trends in Chemical Physics Research. Ed. A. N. Linke. Nova Science Publishers, Inc. Hauppauge, NY. 2005. In press.

In: Progress in Chemical Physics Research
Editor: A. N. Linke, pp. 115-120
ISBN: 1-59454-451-4


*Chapter 6*

# PARADOXES OF THERMAL EXPANSION

***I. A. Stepanov***[1*]
Institute of Physical Chemistry, University of Freiburg, Albertstrasse 23a, Freiburg 79104, Germany

## Abstract

It is shown that the dependence of negative thermal expansion coefficient of many substances on the temperature contradicts to an important thermodynamic relation $(\partial C_P/\partial P)_T = -TV((\partial\alpha/\partial T)_P + \alpha^2)$. It is supposed that there are oscillations at the $C_P(T)$ and $\alpha(T)$ curves at $\alpha > 0$ for some substances and they are in reverse phases.

**Keywords:** negative thermal expansion coefficient, isobaric heat capacity, the first law of thermodynamics, $CeAl_3$, Si, UNiGe, UPt, supercooled water

In [1-11] dependence of the thermal expansion coefficient $\alpha$ and heat capacity $C_P$ on the temperature has been obtained for a number of substances. There is a thermodynamic relation using which one can verify the first law of thermodynamics [12–13]:

$$(\partial C_P/\partial P)_T = -TV((\partial\alpha/\partial T)_P + \alpha^2). \quad (1)$$

One can show that the dependence $\alpha(T)$ found in [1-11] contradicts to the relation (1) for negative $\alpha$. In all temperature intervals under the scope of this paper, $\alpha^2 << |(\partial\alpha/\partial T)_P|$ and can be omitted. In [14] this contradiction has been shown for $ZrW_2O_8$ and $HfW_2O_8$, and in [15] an analogous contradiction to thermodynamics has been shown for iodine crystals under pressure.

In [1-5] $\alpha(T)$ and $C_P(T)$ have been obtained for $CeAl_3$: Figures 1 – 5. From them one can deduce that $\alpha < 0$, $(\partial\alpha/\partial T)_P < 0$ and $(\partial C_P/\partial T)_P > 0$ at $0 < T < 0.4$ K. In [14] it has been proven that

[1] Permanent address: Juglas 1a-20, Riga, LV-1024, Latvia.
* E-mail address: istepanov@email.com

$$\text{sign } (\partial C_P/\partial P)_T = \text{sign } (\partial C_P/\partial T)_P \cdot \text{sign } \alpha. \quad (2)$$

Therefore, the left part of (1) and (2), $(\partial C_P/\partial P)_T$ is less than 0 in this interval. Thus both sides of (1) have different signs. In [3] dependence $\alpha(T)$ has been found also in the pressure range 0.4 – 8.2 kbar. In this range $CeAl_3$ has no negative expansion.

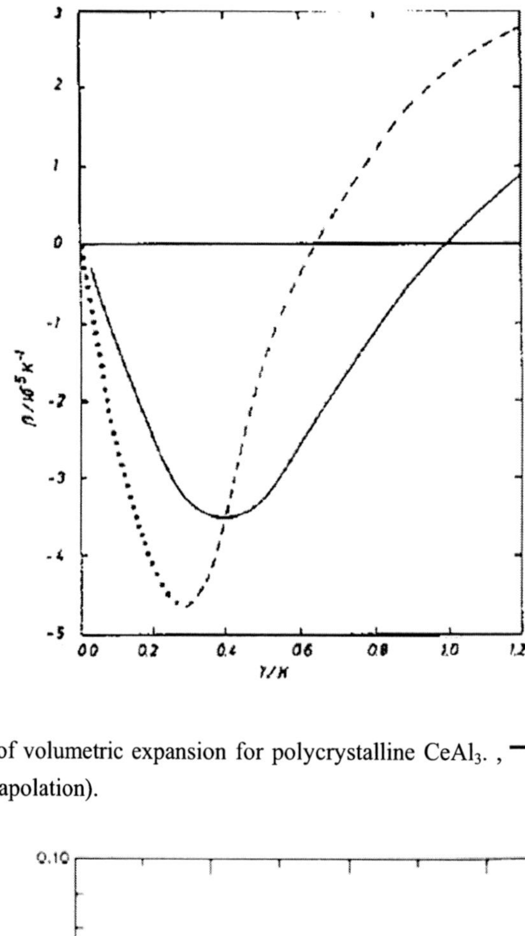

Figure 1. Coefficient of volumetric expansion for polycrystalline $CeAl_3$. , ———————, [1], - - - - - -, [2], · · · · · · ·, [3] (extrapolation).

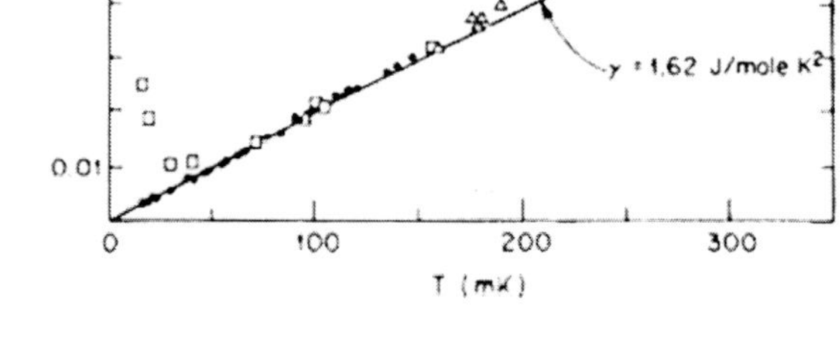


Figure 2. Specific heat of $CeAl_3$ at very low temperatures in zero field (•, Δ) and in 10 kOe (□) [2].

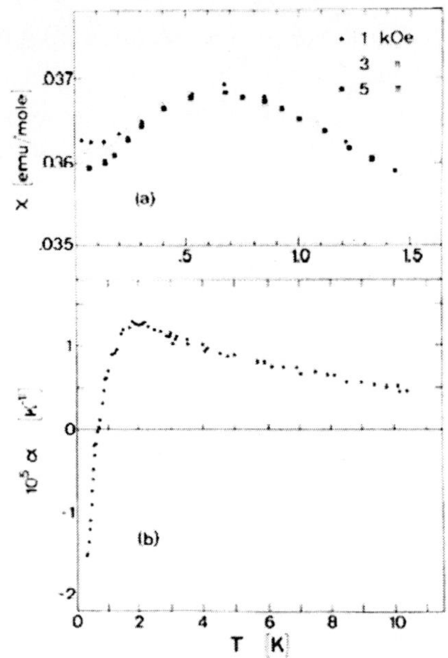


Figure 3. (a) Susceptibility of polycrystalline $CeAl_3$ in different magnetic fields below 1.5 K. (b) Linear thermal expansion coefficient of a polycrystalline sample of $CeAl_3$ below 10 K [2].

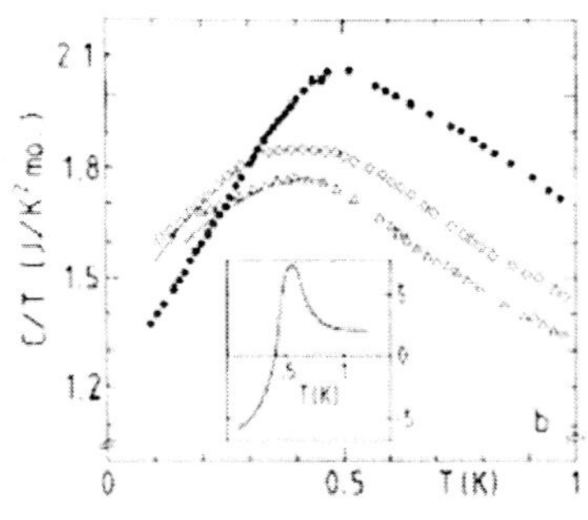


Figure 4. C/T vs T for $CeAl_3$ as a function of magnetic field (•, 0 Tesla, ◊, 1 Tesla, Δ, 4 Tesla). The maximum experimental error is indicated by the symbol size. The inset shows ∂ln(C/T)/∂lnV vs T [4].

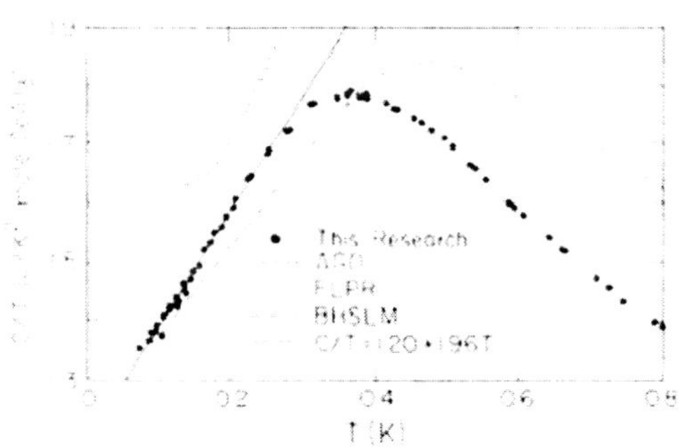


Figure 5. Specific heat of $CeAl_3$ for T ≤ 0.8 K and P = 0. Curves representing data from three other laboratories are also shown [5].

At T > 0.4, $\alpha < 0$, $(\partial\alpha/\partial T)_P > 0$ and $(\partial C_P/\partial T)_P < 0$. From (2) it follows that $(\partial C_P/\partial P)_T > 0$. Again both parts of (1) have different signs.

In [6] dependences $\alpha(T)$ and $C_P(T)$ were given for Si (Figures 6 and 7). It is clear that the data contradict to (1) at T < 0 < 85 K. It is possible that the curve $C_P(T)$ decreases from T ≈ 85 K till T ≈ 120 K (the point of intersection of $\alpha(T)$ with the line $\alpha = 0$) but the experimenters omitted the points. From T ≈ 120 K it begins again to increase.

At T > 120 K, $\alpha > 0$, $(\partial\alpha/\partial T)_P > 0$ and $(\partial C_P/\partial P)_T > 0$ according to (2). There is again a contradiction to (1). One can suppose that there are oscillations at the $\alpha(T)$ and $C_P(T)$ curves and they are in reverse phases. One can detect them if to improve the resolution. In [7] a peak has been detected at the $C_P(T)$ curve at T ≈ 4 K when resolution was improved. There dependences $C_P(T)$ and V(T) were given for UNiGe. There are positive peaks at the $C_P(T)$ curve and negative peaks at the $\alpha(T)$ curve at the same T. It contradicts to (1). The same situation is in [8] and there one can see large maximum and minimum (at the same temperature) at the $\alpha(T)$ and $C_P(T)$ curves of UPt, respectively, and they are in reverse phases.

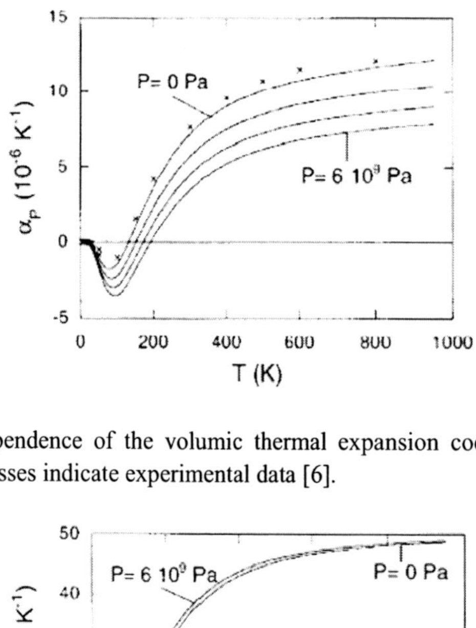


Figure 6. Temperature dependence of the volumic thermal expansion coefficient $\alpha_P$ of Si for four different pressures. The crosses indicate experimental data [6].

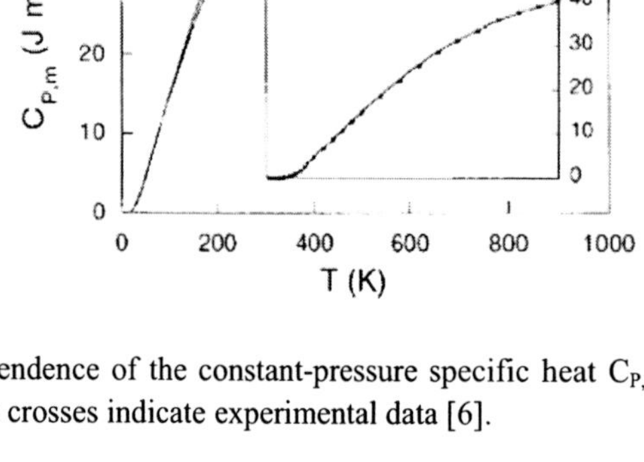


Figure 7. Temperature dependence of the constant-pressure specific heat $C_{P,\,m}$ of Si for two different pressures (calculation). The crosses indicate experimental data [6].

In [9-11, 16] it is shown that negative expansivity of supercooled water gets more and more negative as temperature approaches 228 starting from 273 K. At the same time, $C_P(T)$ increases. As it has been shown above and in [16], this behaviour contradicts to (1).

A possible explanation of these phenomena can be found in [13-18]. There it has been supposed that for substances with negative thermal expansion, the law of conservation of energy must have the following form: $\delta Q = dU - PdV$. (In [13] there is a misprint in equation (11). The correct equation is: for $\alpha < 0$, $C_V = C_P k_S/k_T$). If to derive (1) using this formula, one obtains

$$(\partial C_P/\partial P)_T = TV((\partial\alpha/\partial T)_P + \alpha^2). \quad (3)$$

Notice the change of sign with respect to (1).

## References

[1] Ribault M., Benoit A., Flouquet J. and Palleau J. (1979) Thermal expansion of $Al_3Ce$ at very low temperatures. *J. Physiq. Lettr.* **40**, 15, L413-417

[2] Andres K., Graebner J.E. and Ott H.R. (1975) 4f-virtual-bound-state formation in $CeAl_3$ at low temperatures. *Phys. Rev. Lett.* **35**, 26, 1779-1782

[3] Barron T.H.K. (1992) Pressure dependence at low temperatures of thermodynamic properties of the heavy fermion compound cerium-aluminum ($CeAl_3$). *J. Phys.: Condens. Matter* **4**, 35, L455-L460

[4] Bredl CD., Horn S., Steglich F., Lütthi B. and Martin R.M. (1984) Low-temperature specific heat of $CeCu_2Si_2$ and $CeAl_3$: coherence effects in Kondo lattice systems. *Phys. Rev. Lett.* **52**, 22, 1982-1985

[5] Brodale G.E., Fisher R.A. and Phillips N.E. (1986) Pressure dependence of the low-temperature specific heat of the heavy-fermion compound $CeAl_3$. *Phys. Rev. Lett.* **56**, 4, 390-393

[6] Rignanese G.-M., Michenaud J.-P. and Gonze X. (1996) Ab initio study of the volume dependence of dynamic and thermodynamic properties of silicon. *Phys. Rev. B: Condens. Matter* **53**, 8, 4488-4497

[7] Dmitriev V.M., Eropkin V.N., Gurevich A.M., Isakina A.P., Kobelev V.M., Prokhvatilov A.I., Shlyk L.V., Troc R. and Tran V.H. (1991) On anomalous behavior of heat capacity and structural properties of uranium nickel germanium (UNiGe) at temperatures below 100 K. *Sov. J. Low Temp. Phys.* **17**, 9, 547-548

[8] Kamma C.F., Frings P.H., Klaasse J.C.P. and Franse J.J.M. (1983) Thermal expansion and specific heat anomalies in uranium-platinum below the ferromagnetic ordering temperature. *Physica B+C* (Amsterdam) **119,** 1-2, 72-77

[9] Henn A.R. and Kauzmann W. (1989) Equation of state of a random network, continuum model of liquid water. *J. Phys. Chem.* **93**, 9, 3770-3783

[10] Angell C.A. (1983) Supercooled water. *Ann. Rev. Phys. Chem.* **34**, 593-630

[11] Angell C.A. (1976) Thermodynamic properties of bulk and occluded supercooled water and ionic solutions. Tech. Rep. Purdue Univ. Water Resour. Res. Cent., 75, 15 pp.

[12] Guemez J., Fiolhais C. and Fiolhais M. (1999) Thermodynamics at work: The pressure derivative of the specific heat. *Am. J. Phys.* **67**, 1100-1104

[13] Stepanov I.A. (2000) Thermodynamics of substances with negative thermal expansion coefficient. *Computer Modelling & New Technologies* **4**, No 2, 72-74. (available at www.tsi.lv)

[14] Stepanov I.A. (2005) Negative thermal expansion of $ZrW_2O_8$ and $HfW_2O_8$: contradiction to a thermodynamic relation. – Progress in Chemical Physics Research. Ed. A. N. Linke. Nova Science Publishers, Inc., Hauppauge, NY, pp. 255-262

[15] Stepanov I.A. (2004) Thermodynamics of iodine under high pressure at low temperatures. *Physica B* **349**, 1-4, 251-253

[16] Stepanov I.A. (2006) Supercooled Water: Contradiction to a Thermodynamic Relation. - Progress in Chemical Physics Research. Nova Science Publishers, Inc. Hauppauge, NY. Accepted

[17] Stepanov I. A. (2005) $(\partial S/\partial V)_U$ is negative for substances with negative thermal expansion. - Trends in Chemical Physics Research. Ed. A. N. Linke. Nova Science Publishers, Inc., Hauppauge, NY, 2005. In press

[18] Stepanov I. A. (2005) The Difference Between the Heats of Chemical Reactions Obtained by the Van't-Hoff Equation and by Calorimetry. - Trends in Chemical Physics Research. Ed. A. N. Linke. Nova Science Publishers, Inc. Hauppauge, NY. 2005. In press

In: Progress in Chemical Physics Research
Editor: A. N. Linke, pp. 121-135

ISBN: 1-59454-451-4


*Chapter 7*

# INFRARED SPECTROSCOPY PLAUSIBLE [13]

***Petra Schulz***
Theodor-Francke-Weg 65, D-38116 Braunschweig,
Germany

## 1 Introduction

In the infrared spectroscopy (rotation-vibration spectroscopy) molecules absorb photons in the energy region of the heat energy. For a certain kind of molecule is to be observed thereby a line-rich spectrum. The quantitative treatment of spectroscopic data is well-known owing to mathematical models on the basis of quantum mechanics into all details. The qualitative interpretation remains however mysterious, although we live in the age of the nano-technology. In the present work a qualitative interpretation attempt is undertaken. This attempt of explanation is e.g. important, if one would like to understand the dispersion at individual molecules or also at many particles in a lattice structure by coherent laser bombardment.

## 2 What are Photons?

### 2.1 The Photon of the Free Electron

Large question now: What are photons actually? At the DPG conference in Dresden I tried, to define mass, charge and spin plausibly and used thereby a descriptive photon model [7]. Assumed, it shall be shot on electrons with photons. Then the simplest photons (approximately as impact partners of resting, orbital angular momentumless electrons) are most properly comparable with projectile balls from a rifle. Photons have thus

a) a forward urge, an orbital angular moment, (in fig. 1 represented as horizontal arrow). Photons are however faster than rifle bullets, because they fly with speed of light. I do not want to deal in the following with the orbital angular moment, which may be constituted however, but instead of this with the spin of the light particles.

b) Photons swivel like insidious footballs or billard balls, thus they rotate around their own axis, they possess a spin, (in fig. 1 represented as curved arrow). The spin is in the simplest case either in or opposite to the direction of propagation - thus a longitudinal polarized photon (s. fig. 1a) - or perpendicularly to the direction of propagation - thus a transverse polarized photon (s. fig. 1b).

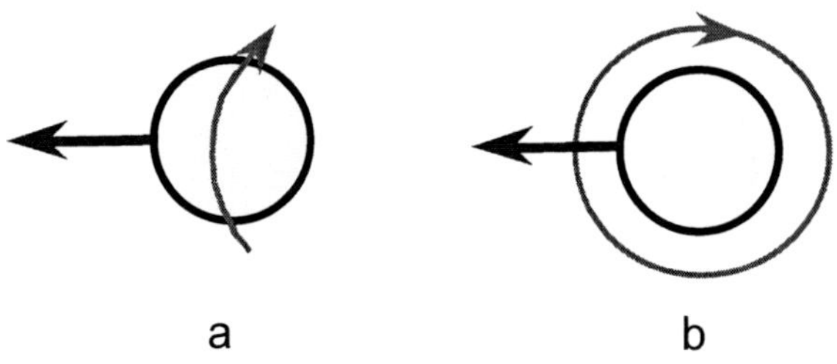


Fig. 1: Photon with forward momentum and spin. a) example of a longitudinal polarized photon; b) example of a transverse polarized photon

In the Maxwell theory only the transverse polarized photons are considered. That is certainly an inadequacy, at least in the (molecular) spectroscopy. Linear combinations of the two polarizing conditions in the arbitrary relationship occur of course at the same time too.

## 2.2 The Photons of the Molecules

The descriptive idea just introduced about photons shall be in principle also maintained, if the molecules are excited by a photon as for example in the infrared spectroscopy. I would like to see the changes in the molecule qualitatively descripted and not abstractly by the wave function. In the written version of my DPG lecture in Bremen [8] I defined the photons completely rough as the negative of the particles. (That should be only one additional statement, why one needs the group theory in the spectroscopy.) Perhaps I should specify still more: The absorbed or also later re-emitted photons must be just as structured as the stimulated or the calmed down parts in the particles. Thus, at the impacts I do not need to mind the structure function (scattering matrix) of the particles.

An infrared spectrometer shoots levogyric and/or dextrogyric photons in and transverse to the direction of propagation of the photons at the molecules. A photon can get rid of its rotational energy, if it hits at a molecule. All molecules are potential tops.

## 2.3 Permeability of the Particles for Photons

Photons can pass at impacts through particles. I still remind thereby at the "toc toc toy", the pendulum chain, and/or also at the relay race of the photons by the electrons at the electric conduction in metals. If one refrains from the first and last particle, only photons are moved. Photons are permeable for many particles. This transparency of the particles for photons depends thereby particularly on the respective energy state of the particles.

### 2.4 The Wave Function of the Photon

I know: the wave function of the single particle is exactly like the refractive index [8] a real quantity. And I extrapolate with good right: also the wave function of an ensemble of many particles is real during the interaction with photons under warranty.

# 3 Rotation Spectroscopy

### 3.1 Developing a Rotation Spectrum

At diatomic (rod-shaped) molecules no rotation is observed around the body axis by spectroscopic methods, but (in the suitable case with dipole moment-afflicted molecules) only around the perpendicular sense of the connection axis of the two atoms. Therefore we want to turn from the beginning into polyatomic molecules. A top molecule with three-fold axis is drawn in fig. 2. That can be e.g. methyl chloride ($CH_3Cl$) or perhaps also silyl chloride ($H_3SiCl$). Such a molecule resembles a child gyroscope, which stands on the head, its point. Assumed, two photons fly on the molecule: Photon p (p for parallel) and photon v (v for vertical). Because of the clarity in fig. 2 the structure of the photon was suppressed.

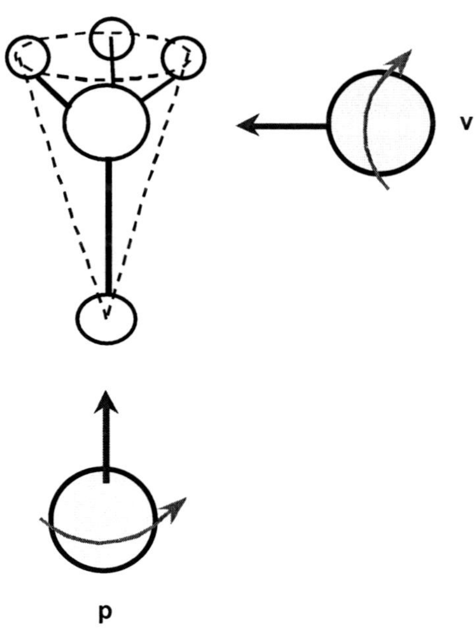


Fig. 2: Photon p flies on the symmetric top molecule $CH_3Cl$ or $H_3SiCl$ from the head - parallel - to the body axis (as when rotating a child gyroscope) and photon v vertically to it.

If the photon v hits such gyroscope on ventral side, then the top spins perpendicularly to its body axis, therefore I name this movement as perpendicular rotation. If photon p approaches from the head, it will let the top rotate around its symmetry axis, we know this movement from the child gyroscope. Because the rotation axis (body axis) lies parallel (even directly in the body axis), I name the taking place movement parallel rotation, s. fig. 2. Perpendicular and parallel rotation are by the way no usual designations. In principle there are two perpendicular rotations, the drawn in and in addition the perpendicular one. In the case of symmetric top molecules the two perpendicular rotations are degenerated (energetically equivalent).

The following assumption appears justified: The mentioned procedures occur not directly at first attempt, because they need a certain starting time. Molecule and photon must practice these regularly coordinated movements only. But several beginnings or rework are necessary. Fig. 3a shows in simplified manner a parallel rotation and fig. 4a a perpendicular rotation. For simplification the top molecules were only represented in the side view as triangles. In fig. 3b and 4b after further simplification only the projection of the rotation circulation from the side is drawn.

For the parallel rotation the quantum number *K*, called also projection quantum number was introduced. For the total angular momentum one uses the total angular momentum quantum number *J*, s. fig. 4. (The total angular momentum is frequently diagonal arranged to the symmetry axis of the molecule.) For the perpendicular rotation is (*J* - *K*) the associated quantum number combination. (*K* is called projection quantum number, because *K* belongs to the projection of the total angular momentum on the body axis.) Parallel and perpendicular rotation can take place at the same time. By the way the conical movement of the body axis is called precession.

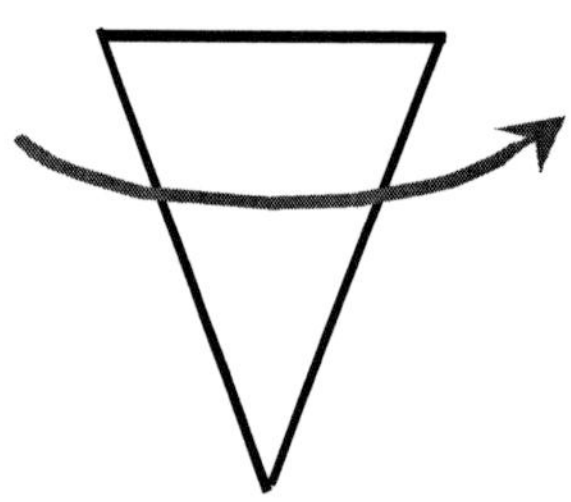

Fig. 3a: Parallel rotation

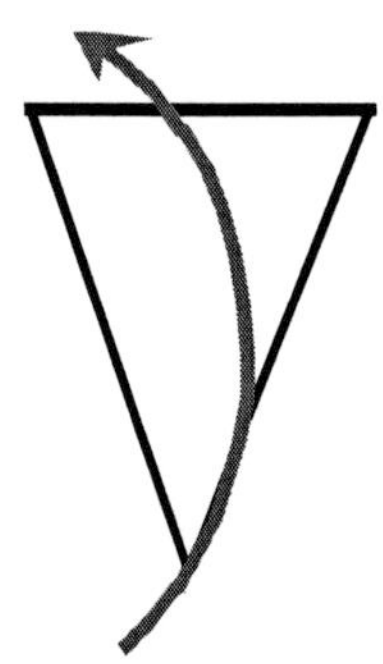

Fig 4a: Perpendicular rotation

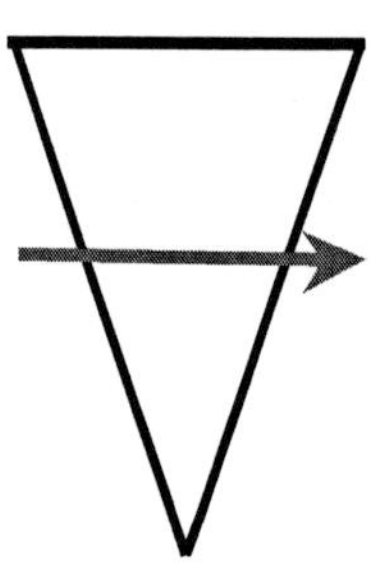

Fig. 3b: Parallel rotation

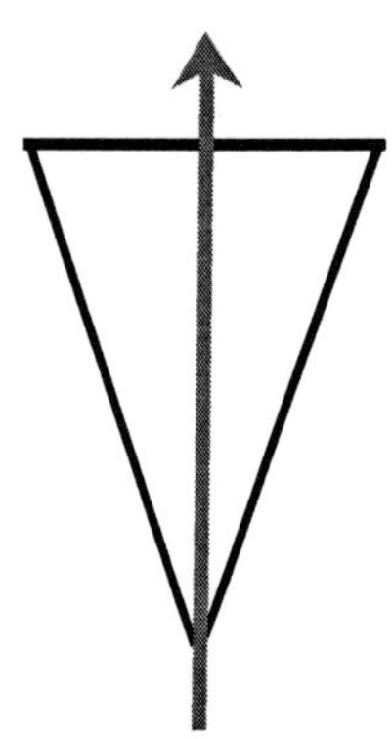

Fig 4b: Perpendicular rotation

## 3.2 Appearance of a Rotation Spectrum

Rotation spectra of diatomic molecules are particularly simple. That is because these molecules can implement allegedly only perpendicular rotations and no parallel rotation around the stick axis (nuclear connection axis), s. also chapter 5.6. Therefore the rotation spectrum of gaseous hydrogen chloride, HCl, is shown in the fig. 5 as an example of a perpendicular rotation. The resolution is so bad in such a favorable manner, that the lines of the two isotopomers with $^{35}Cl$ and $^{37}Cl$ are not resolved. A rough fast explanation to the rotation spectrum is anticipated: With increasing wavenumber the molecule loads one step more at perpendicular rotation.

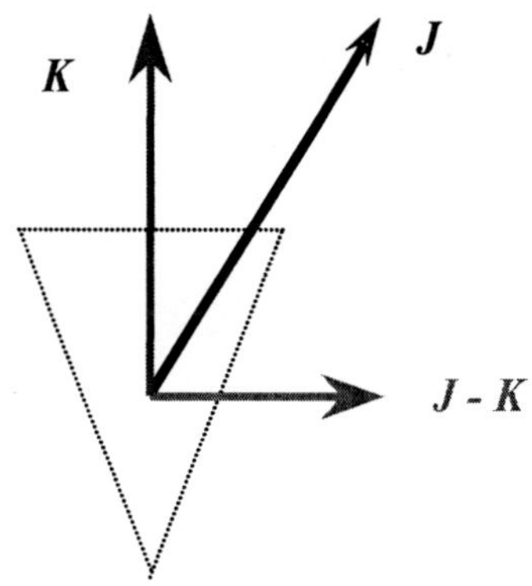


Fig. 4: The three quantum numbers for the rotation of a symmetric top: *J* stands for the total angular momentum (in arbitrary situation to the molecule), *K* for the parallel rotation (projection of the associated vector of *J* on the body axis), and (*J* - *K*) for the perpendicular rotation.

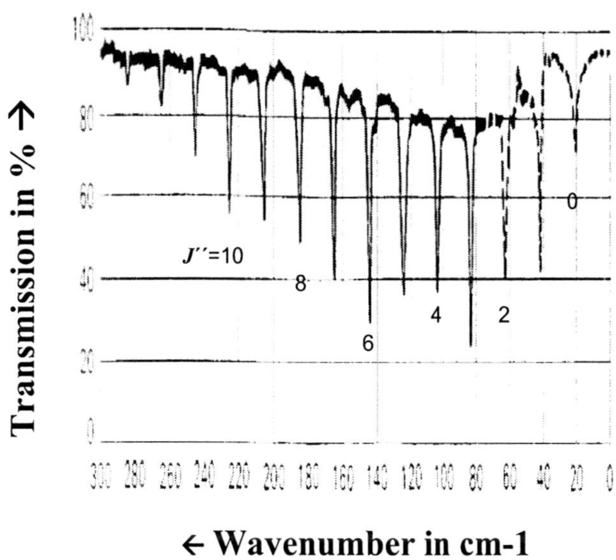


Fig. 5: Pure rotation spectrum of gaseous HCl (from DEMUTH/KOBER [3], slightly changed). The broken lines with the rotation quantum number in the initial state $J''$ = 0, 1 and 2 were simulated.

# 4 Rotation-Vibration Spectroscopy

The molecules can rotate and in addition vibrate also. An oscillation consists of two opposite rotations. These two rotations moving in opposite directions for an oscillation quantum are accordingly high-energetic in the rotation-vibration spectroscopy (rovibrational spectroscopy).

The oscillations take place in the rotation-vibration spectroscopy either perpendicularly to the body axis (perpendicular vibration or **perpendicular band**, s. fig. 6a) and/or parallel (exactly taken in) the body axis (parallel vibration or **parallel band**, s. fig. 6b). Those are by the way quite usual designations (s. e. g. [9]). I quote by the way consciously the book of STUART, because this author tried to bring for some spectroscopic procedures a classical explanation. With inclined bombardment of the photon on the particle can be stimulated both the perpendicular and the parallel band, if the energy is sufficient. An example of a parallel band is discussed mathematically in the book of DEMUTH and KOBER [3] for progressive students on school level (however rather probably only for male pupils).

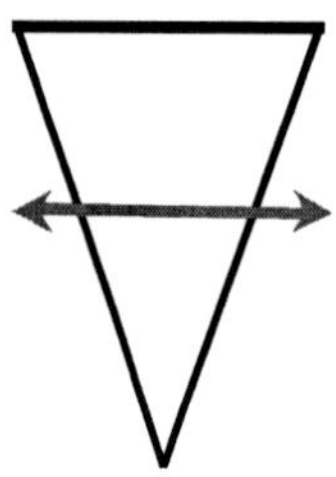

Fig. 6a:
Perpendicular vibration

Fig. 6b:
Parallel vibration

The vibration quantum number is called $v$. In the absorption spectroscopy is $\Delta v = +1$ and in the emission spectroscopy $\Delta v = -1$. This selection rule will be repeated again in chapter 4.2.

## 4.1 Term Symbolism of the Rotation-Vibration Spectroscopy

We became acquainted with the most important quantum numbers of the rotation-vibration spectroscopy. They are sufficient already to characterize (indicate) a line in the spectrum, a peak. The following term symbolism is internationally usual:

$$^{\Delta K}\Delta J_{K''}(J'')$$

$\Delta$ means final state minus initial state. For the absorption spectroscopy this means: stimulated minus ground state. $K''$ and $J''$ are the quantum numbers in the ground state, normally without vibration stimulation. $\Delta K$ and $\Delta J$ are replaced by capital letters. (Please don't be afraid! The term symbolism appears somewhat annoying. But for the researchers, who have to do frequently with rotation-vibration spectroscopy, this symbolism quite means a facilitation.)

$\Delta K$ =-1 or $\Delta J$ =-1: P

$\Delta K$ = 0 or $\Delta J$ = 0: Q

$\Delta K$ = 1 or $\Delta J$ = 1: R

For parallel bands the symbol Q for $\Delta K = 0$ is usually omitted.

## 4.2 Selection Rules

Because of completeness here are a few words to the selection rules. To **rotation spectra** applies

$$\Delta J = \pm 1,$$

whereby $\Delta J = +1$ (thus the absorption) is probably the most frequently observed case.

In the **rotation-vibration spectroscopy** the oscillation quantum number $v$ is important because of the vibration:

$$\Delta v = \pm 1.$$

In addition the quantum number $K$ for the rotation around the body axis plays a role frequently, not for **parallel bands**, because for these obey

$$\Delta K = 0,$$

but **perpendicular bands** apply of the selection rule

$$\Delta K = \pm 1$$

and

$$\Delta J = \pm 1$$

as well as additionally also still

$$\Delta J = 0.$$

Diatomic molecules supply with only parallel bands and molecules with more atoms additionally to the parallel bands also still perpendicular bands.

Perpendicular bands are a little more complicated and not suitable for a treatment in school because of the line variety. I will therefore touch this topic only at the edge (because of completeness).

Table 1: Some important selection rules for symmetric tops.

<table>
<tr><td colspan="2">rotation spectroscopy<br>$\Delta v$=0; $\Delta J$=±1</td></tr>
<tr><td colspan="2">rotation-vibration spectroscopy<br>$\Delta v$=±1</td></tr>
<tr><td>parallel band<br>$\Delta K$ =0; $\Delta J$=(0),±1</td><td>perpendicular band<br>$\Delta K$ =±1; $\Delta J$=0,±1</td></tr>
</table>

About the selection rule $\Delta J = 0$ for a parallel band s. briefly chapters 5 and 5.4.

## 4.3 Appearance of Rotation-Vibration Spectra

In the following two illustrations one experimental example each of a parallel band of HCl is shown (in fig. 7) and a perpendicular band of $CH_3Cl$ (in fig. 8). The R branch in the rotation-vibration spectrum of HCl in fig. 7 resembles from the appearance, if one abstracts from the resolution, the pure rotation spectrum of HCl in fig. 5. That generally applies to the R branches within the parallel bands of diatomic molecules up to symmetric top molecules. In contrast to the rotation spectrum of the fig. 5 the resolution of the rotation-vibration spectrum in fig. 7 is so good that the peaks of the two different heavy isotopomers with $^{35}Cl$ and $^{37}Cl$ are resolved. Compare additionally also still the schematic illustration in fig. 9, the first row with $K = 0$, that is typical for the rotation-vibration spectrum of a diatomic molecule.

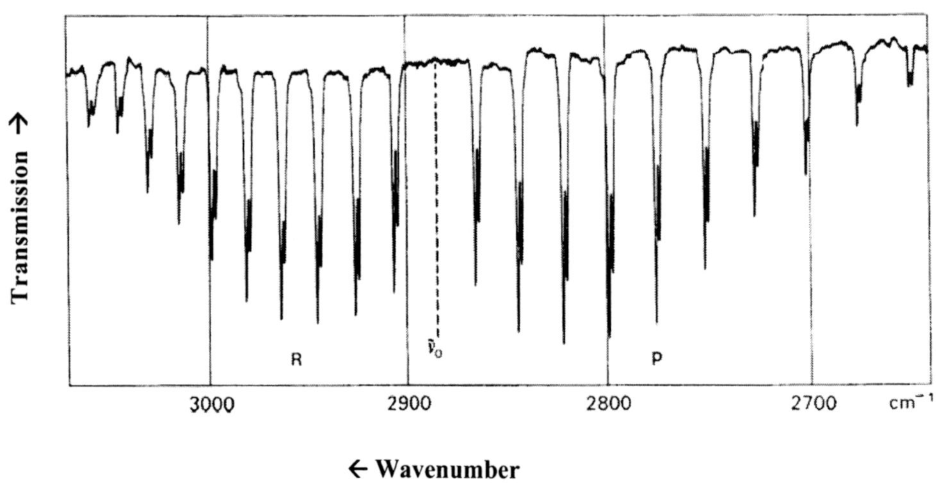


Fig. 7: Example of a parallel band is the rotation-vibration spectrum of HCl (from DEMUTH/KOBER [3])

Because of the weak resolution at the perpendicular band of $CH_3Cl$ in fig. 8 one can recognize only the $^{R}Q$ and $^{P}Q$ branches. That has the following reason: The lines with different $J$, belonging to a $K$ value, are nearly at the same place on the wave number axis and add their intensity. The remaining lines go under regarding the noise.

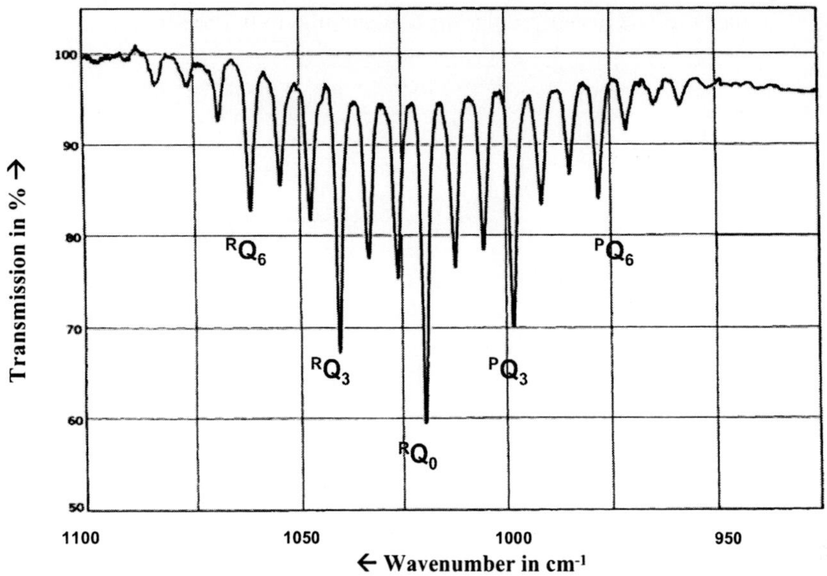


Fig. 8: Example of a perpendicular band is the rotation-vibration spectrum of $CH_3Cl$ at weak resolution (from BARROW [1], slightly varied)

# 5 Interpretation of Spectroscopic Transitions

What does one recognize the oscillation? One does not see the vibration directly in the spectrum. It can be suspected only from the position of the peaks. In fig. 7 the band origin of the vibration is marked as a dashed line. One sees rotation transitions however. They are very similar to the rotation-, the rotation-vibration- and the electronic vibration-rotation-spectra (if the latter are well resolved).

Here is now betrayed, which changes of rotation arise at the individual branches of a parallel band ($^QQ$ , $^QR$ , $^QP$ branch) and a perpendicular band ($^RR$ , $^PP$ , $^RQ$ , $^PQ$ , $^RP$ , $^PR$ branch) in detail, because this is not to find in any textbook, s. table 2 and table 3.

Table 2: What happens at the parallel band in addition to the oscillation?

<table>
<tr><td colspan="2">$^QQ$<br>(Parallel and perpendicular) rotation condition remains constant</td></tr>
<tr><td>$^QR$<br>Perpendicular rotation to load</td><td>$^QP$<br>Perpendicular rotation to unload</td></tr>
</table>

Whether at a parallel band a $^QQ$ branch is visible, on it decide the nuclear spins in the molecules and thus the nuclear spin statistics. But this topic is not to be deepened here.

Table 3: What happens at a perpendicular band in addition to the oscillation?

| $^{R}R$ Parallel rotation to load | $^{P}P$ Parallel rotation to unload |
|---|---|
| $^{R}Q$ Perpendicular into parallel rotation to convert | $^{P}Q$ Parallel into perpendicular rotation to convert |
| $^{R}P$ Perpendicular into parallel rotation to convert and perpendicular rotation to unload | $^{P}R$ Parallel into perpendicular rotation to convert and perpendicular rotation to load |

## 5.1 Concerning the Position of the First Lines

In the following for simplification is set $K = K''$. The spectra of symmetric top molecules are regarded. Where now do the first observable lines lie? The first lines of a parallel band starting from $K = 0$ are:

$$^{Q}R_{K}(K)$$

$$^{Q}Q_{K}(K)\ [^{Q}Q_{K}(K) \text{ only starting from } K=1]$$

$$^{Q}P_{K}(K+1).$$

(Caution, the expressions in the round parentheses indicate for a given value of $K$ the smallest permitted value for $J$.) As first lines of a perpendicular band in the P branch are permitted those starting from $K$=1:

$$^{P}R_{K}(K)$$

$$^{P}Q_{K}(K)$$

$$^{P}P_{K}(K),$$

and in the R branch already starting from $K$=0:

$$^{R}R_{K}(K)$$

$$^{R}Q_{K}(K+1)$$

$$^{R}P_{K}(K+2).$$

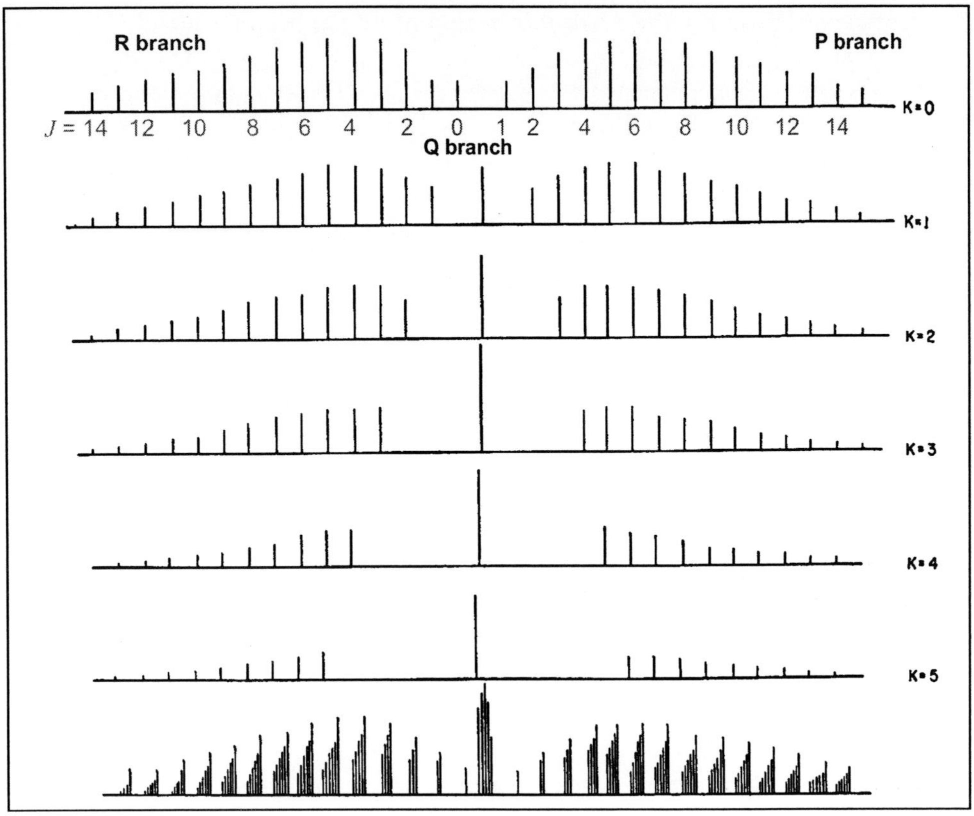


←Wavenumber

Fig. 9: The subbands of a parallel vibration for $K$=0 to $K$=5 with $J$=0 to maximal 15. In the lowest spectrum all subbands mentioned are added (from CONDON [2], easily varied). A Q branch is to be observed only for certain nuclear spin constellations within a symmetric top.

Here only briefly to the better comprehension an example: The first lines of a parallel band are for $K$ = 0: $^{Q}R_0(0)$ and $^{Q}P_0(1)$, s. fig. 9 in the first row. Even if no perpendicular rotation is present as with $J$ = 0, such rotation can be loaded: That corresponds to the transition $^{Q}R_0(0)$. A perpendicular rotation can be delivered only, if such exists, thus starting from $J$ = 1, this is the transition $^{Q}P_0(1)$.

For advanced readers still another similar addition from the Raman spectroscopy: For $K$ = 0 there is no emission line (anti-Stokes line) at the absolute zero, but only an absorption line (Stokes line).

A parallel band can be already very line-rich e.g. at a good resolution, s. fig. 9, last line. One can bring a certain order and simplification into this line variety, if one sums parts of the spectrum in respect of $K$ values in the initial state. These parts are called subbands and consist at least of a P and R branch. Such a spectrum can be divided into subbands for different $K$ values of a polyatomic molecule. But at a diatomic molecule only the subband with $K$=0 can be observed.

## 5.2 Concerning the Possible Interpretation of Oscillations

A vibration is conceivable in principle by a strong rotation, with which in addition an equally strong rotation in opposite direction occurs. That was already mentioned in chapter 4. I do not want to deal with the quite interesting cases, at which spins or weak rotations waive themselves and a weak vibration results. The following two sentences are meant only for particularly advanced persons: For the infrared spectroscopy that means that one needs slightly less energy for the next higher vibration stimulation (anharmonicity). In elementary particle physics vibration excitations supply extra contributions to the form factor of a nucleus [10].

## 5.3 Concerning the Manifold of the Molecules

The search for the real total wave function is a difficult puzzle. A problem represents the presence of very many molecules and to their interactions within a large observation period. An individual molecule supplies only a small peak at photon bombardment. I repeat again: If the wave function of an individual molecule is real, naturally the total wave function of the particle ensemble must be also real.

Because of the large observation period we see the total output of many molecules and their isotopomers simultaneously, thus a lot of peaks. But we do not see, which transitions are permitted and how the peaks have arisen. In one point we can be somewhat sure (it applies the comparatibility of the means, speak the energy): Normally an isomer will hardly transform into a differently other one (transmutate), this is only possible in the spectroscopy of high-energy physics. Attention, practically entire elementary particle physics is nothing else but high-energy spectroscopy.

## 5.4 Nuclear Spin

Nuclear spin decides on the existence of a Q branch at a parallel band as well as on the intensity relationship of the lines in a subband of a parallel or a perpendicular vibration. Therefore the lines with $K$ values, which are divisible by 3, are twice as strong like the remaining lines e.g. of the molecules with a triple axis of rotation such as $CH_3Cl$ (s. fig. 8). One finds generally understandable references to the nearly same topic in the textbooks for chemists in the chapters about the nuclear magnetic resonance spectroscopy (NMR spectroscopy) at hydrogen cores.

In the mathematical modelling of the rotation-vibration spectroscopy nowadays the nuclear spin quantum number plays still no role, since the absolute resolution of the spectra is not yet as fine as in the NMR spectroscopy.

## 5.5 Gravitation

Who likes it, can file the lines of a spectrum in the chapter quantum gravitation and is completely well informed thereby about the nature of physical procedures for the first.

## 5.6 The "Sacred Cow" of the Selection Rules

The validity of the selection rules depends relatively strong on the test conditions. The chapter selection rules is qualitatively not at all understood unfortunately in the textbooks and therefore only toilsomely comprehensible in accordance with the statements of the quantum theory. The attempt of an explanation by HERRMANN [4] is praise worth, but unfortunately nevertheless too simple. I can supply however only a few additions:

No rotational spectrum is observed by atoms. Of course this does not mean that the atoms do not rotate. Since they do not possess any dipole moment, they cannot supply any rotational spectrum. I suggest the following explanation possibility: Spherical-shaped corpuscles such as atoms (or even electrons) load in the observation period precisely as many rotational energy as they unload so that in total no line can be observed in the spectrum, at least in the field-free to field poor space.

It seems to be like this that the examined spectroscopic systems ignore especially then a selection rule whenever they are far distant from the thermal equilibrium.

Thus test conditions of astrophysics permit to observe forbidden lines by the extremely weak particle number density (no impacts between the particles) and the enormous radiation path. Under terrestrial laboratory conditions highly symmetrical particles, electrons or molecules make elastic impacts among themselves, so that at the detector no changes are registered. Asymmetrical uniform molecules may be able to execute still partial elastic impacts under themselves. But at the walls the excited molecules do not get rid of their excess energy because of the mismatching figure of the wall molecules, so that by these asymmetrical molecules one peak can be registered.

Furthermore a high intensity can ensure during short time of measurement that forbidden lines appear in the spectrum.

Also adding of foreign particles can cause miracle (whereupon e.g. the operational mode of the helium-neon-lasers is based).

If in the observation period just as many molecules are excited as calmed down, then one will see no peak at all in the spectrum, that is so to say a direct consequence of the selection rules. That applies not only to the atoms mentioned above but also to symmetrical molecules, e.g. the vibration and/or rotation of homonuclear diatomic molecules.

To the selection rule $|\Delta J| > 1$ if changes of rotation happen: At the DPG conference in Berlin 1997 [6] I referred the observability of forbidden lines from more recent time at silyl halides with $\Delta K = \pm 3$ [11], [12]. In the meantime a further selection rule was already observed, namely $\Delta K = \pm 6$ [5]. Despite of these new discovered selection rules the spectroscopy is not revolutionized at present yet.

Here I would like to express still another relevant assumption: Nuclear spin transitions and rotation transitions need comparatively less energy in contrast to the vibration stimulation. Therefore nuclear spin and rotations are stimulated up to much higher quantum numbers, than one believed so far. The rotating and oscillating photons are usually transferred by nearly elastic impacts from particle to particle. This kind of the impacts is observable in the spectrum, because the particle 1, loaded with the photon, hits another particle, which is very probably rather in another nuclear spin and/or rotation condition very probably as the particle 1 unloaded meanwhile trough particle 2. I believe, what I wrote evenly so lapidary there, seems to be for the observation from spectra completely important.

Better resolutions, faster computers will bring out still further new selection rules in the future and will provide for striking surprises. The better spectroscopic transitions will be interpreted and appropriate consciousness for it is present, the more the determination within the micro-cosmic range will find again an authorization into the physics. Up to this goal however still another far cumbersome path is to be taken.

# 6 Conclusion

How can we explain ourselves the formation of an absorption line in principle in the infrared spectroscopy? The fact is undisputed that a photon shifts an angular momentum upon the molecule. At the same time in the rotation-vibration spectroscopy also still another oscillating momentum is unloaded. In order to be able to observe in the field-free area a spectral line, one needs parts in a molecule, which are swivelling, and which can absorb an angular momentum for the own need from the photon for the change of spin, rotation, nuclear spin. This leads to an intensity increase, if transitions arise at nearly the same wavenumber, and that can be only the reason for the fact that transitions become visible at all.

# References

[1] Barrow, G. M.: *Introduction of Molecular Spectroscopy*. Tokio: McGraw-Hill 1962

[2] Condon, E. U. (editor): *Handbook of Physics*, 2nd ed. New York: McGraw Hill 1967

[3] Demuth, R.; Kober, F.: *Grundlagen der Spektroskopie*. Frankfurt: Diesterweg-Salle 1977

[4] Herrmann, F.: Auswahlregeln, verbotene Übergänge. In: *Physik in der Schule* **38**, (2000), p. 74

[5] Ruland, H.: Dissertation. *Rotations-Schwingungsanalyse symmetrischer und asymmetrischer Kreiselmoleküle nahe dem Kugelkreisel*. Wuppertal 2000

[6] Schulz, P.: Kurioses zum Photon (Strange Behavior of the Photon). In: Tagungsband *Didaktik der Physik, Vorträge*. Berlin: Deutsche Physikalische Gesellschaft 1997 (in German)

[7] Schulz, P.: Plausible Definition von Masse, Ladung und Spin (Plausible Definition of Mass, Charge and Spin. In: *CD zur Frühjahrstagung Didaktik der Physik in der Deutschen Physikalischen Gesellschaft*, Dresden 2000 (in German). 2000: http://www.GruppeDerNeuen.de

[8] Schulz, P.: Plausible Erklärungshinweise gegen die Überlichtgeschwindigkeit (Plausible Explanation Attempts against the Supraluminal Propagation of Light). In: *CD zur Frühjahrstagung Didaktik der Physik in der Deutschen Physikalischen Gesellschaft*, Bremen 2001 (in German). 2003: http://www.GruppeDerNeuen.de (in German, abstract in English)

[9] Stuart, H. A.: *Molekülstruktur*, 3rd ed. Berlin: Springer 1967

[10] Greulich, W. (editor): *Lexikon der Physik*. Heidelberg: Spektrum Akademischer Verlag, 1998

[11] Ceausu, A.; Graner, G.; Bürger, H.; Pracna, P.: A New Rovibrational Analysis of the $\nu_2/\nu_5$ and $2\nu_3$ Bands of Monoisotopic $H_3Si^{79}Br$. In: *Journal of Molecular Spectroscopy* **174**, pp. 237-252 (1995)

[12] Papoušek, D.; Bürger, H.; Rahner, A.; Schulz, P.; Hollenstein, H; Quack, M.: High Resolution Infrared Spectra and Simultaneous Rovibrational Analysis of the $\nu_2$, $\nu_3$, $\nu_5$ and $\nu_6$ Bands of $H_3SiF$. In: *Journal of Molecular Spectroscopy* **195**, pp. 263-280 (1999)

[13] Translated from German, easily changed from: Schulz, P.: Infrarotspektroskopie plausibel. In: *CD zur Frühjahrstagung des Fachverbandes Didaktik derPhysik in der Deutschen Physikalischen Gesellschaft*, Leipzig 2002. http://xxx.lanl.gov/abs/physics /0307062

In: Progress in Chemical Physics Research
Editor: A. N. Linke, pp. 137-188
ISBN 1-59454-451-4


*Chapter 8*

# Crystal Orientation Effects on Electronic and Optical Properties of Strained GaN Semiconductors

***Seoung-Hwan Park****
Department of Photonics and Information Enginnering,
Catholic University of Daegu,
Hayang, Kyeongbuk, Korea
***Shun-Lien Chuang***†
Department of Electrical and Computer Engineering,
University of Illinois at Urbana-Champaign,
1406 West Green Street, Illinois 61801

**Abstract**

Crystal orientation effects on electronic and optical properties of strained GaN semiconductors are reviewed. Analytical expressions for the band gap, wave function, and momentum matrix element of bulk GaN semiconductors with an arbitrary crystal orientation are shown using a block-diagonalized Hamiltonian. Self-consistent (SC) solutions considering the spontaneous (SP) and piezoelectric (PZ) polarizations for valence-band structures of a quantum well (QW) with an arbitrary crystal orientation are obtained by solving the Schrödinger equation for electrons, the $6 \times 6$ Hamiltonian for holes, and Poisson's equation iteratively. The optical gain is investigated using the non-Markovian gain model with many-body effects. The optical gain increases with increasing crystal angle, which is mainly due to the increase of the matrix elements with increasing crystal angle. In particular, the optical gain of the $(10\bar{1}0)$-oriented QW structure is largely enhanced over that of the (0001)-oriented QW structure. This can be explained by the fact that the optical matrix elements near the band edge ($k$=0) of the $(10\bar{1}0)$-oriented QW structure significantly increase compared to those of the

*E-mail address: shpark@cu.ac.kr
†E-mail address:s-chuang@uiuc.edu

(0001)-oriented structure, in addition to the reduction of the effective mass. The threshold current density $J_{th}$ of the $(10\bar{1}0)$-oriented QW structure is expected to be about a half of that of the (0001)-oriented QW structure.

**PACS** Valid PACS appear here
**Keywords** Optical gain, GaN, AlGaN, InGaN, Crystal orientation, many-body effect, polarization

# 1 Introduction

The group III nitrides such as GaN, AlN, and InN have drawn a lot of attention for their potential electronic and optoelectronic device applications such as laser diodes (LDs) in the blue and ultraviolet regions. Nakamura[1] has demonstrated room-temperature continuous-wave operation of wurtzite (WZ) InGaN/AlGaN multi-quantum well (MQW) laser diodes with a lifetime of many thousand hours. In general, the threshold current densities of these lasers are much larger than those of the conventional zinc-blende (ZB) based lasers. In the zinc-blende crystals (e.g., GaAs or InP), a biaxial compressive strain in a QW structure reduces the in-plane effective mass of the top heavy-hole band and the threshold current of a semiconductor laser.[2] On the other hand, the hole effective mass of GaN is significantly larger than that of conventional zinc-blende crystals. Also, the introduction of biaxial strain into the (0001)-plane of a wurtzite GaN-based crystal does not effectively reduce the effective masses in the transverse direction.[3, 4, 5] This is due to the fact that, for a wurtzite layer grown along the (0001) orientation, the strain shifts both the heavy hole (HH) and light hole (LH) bands by almost the same amount and the in-plane effective masses remain almost the same as those in the unstrained case.[6, 7]

The wurtzite GaN-based structures are found to have very large spontaneous (SP) polarization in addition to the large strain-induced piezoelectric (PZ) polarization.[8, 9, 10] Experimentally, several reports indicate that there exist giant electric fields even in unstrained GaN QWs: the high carrier density at the GaN/AlGaN interfaces,[11] the redshift of the photoluminescence (PL) transition energy, and the dramatic increase of the carrier lifetime,[12] due to the PZ and SP polarizations. Leroux *et al.*[13] observed the blueshift of the PL transition energy with decreasing barrier width and attributed it to the SP polarization. Theoretical studies[14, 15, 16, 17] also showed that the valence band structures and the overlap integral of the electron and hole wave functions of GaN/AlGaN quantum well structures are affected significantly by both the PZ and SP polarizations for (0001) wurtzite structures. These lead wurtzite GaN-based QWs of the (0001) crystal orientation to require higher carrier densities to generate the optical gain compared to zinc-blende GaAs-based QWs. Thus, it is essential to reduce the hole effective mass or the internal field in order to decrease the threshold carrier density in wurtzite GaN-based QW lasers.

As an additional parameter for band engineering, the crystal orientation effect on electronic and optical properties in wurtzite GaN-based QW lasers has been studied in order to decrease their threshold carrier density. Domen et al.[18] investigated the band structure and optical gain of bulk $(10\bar{1}0)$ GaN. Niwa et al.[19] obtained the valence-subband structures by using the tight-binding method for wurtzite $(10\bar{1}0)$ QW structures. Yeo et al.[20] studied the valence subband structure and optical gain for wurtzite $(10\bar{1}0)$ QW structures by using the multiband effective-mass theory. These results were compared with those for the (0001)-oriented QW lasers. Ohtoshi *et al.*[21] investigated the optical gains of wurtzite GaN/AlGaN QW lasers for various crystal orientations such as $(10\bar{1}5)$, $(10\bar{1}2)$, and $(20\bar{2}1)$. However, most works on the crystal orientation effect of electronic and optical properties in GaN-based QW lasers have been studied under the free-carrier (FC) model without considering the many-body effects and under the flat-band (FB) model without considering the internal fields due to the strain-induced PZ and SP polarization. The electronic and the optical properties of GaN-based QW lasers can be influenced significantly by the internal fields except the $(10\bar{1}0)$ crystal orientation.[22, 23, 24, 25, 26] The wurtzite GaN-based QW structure with a $(10\bar{1}0)$ crystal orientation is interesting because the internal fields are expected to be zero.[17, 27, 28, 29] Also, it is well known that the many-body effects are significant due to the large exciton binding energy in GaN-based QWs.[30]

In this chapter, we review the crystal orientation effects of electronic and optical properties in wurtzite bulk GaN-based crystals and QW structures with PZ and SP polarization using the multiband effective-mass theory. The SC band structures and wave functions are obtained by solving the Schrödinger equation for electrons, the $6 \times 6$ Hamiltonian for holes, and Poisson's equation iteratively.[2, 6, 15] These results are compared with those of the FB model in the absence of SP and PZ polarization effects.

In Section II, we show the general $6 \times 6$ Hamiltonian for wurtzite crystals with an arbitrary crystal orientation. We also derive analytical expressions for the strain components, the piezoelectric field, eigenenergies, bandgaps, wavefunctions, and optical matrix elements for an arbitrary crystal orientation of the strained wurtzite GaN. The $6 \times 6$ Hamiltonian for the valence bands at the band edge is block-diagonalized into two $3\times3$ Hamiltonians, which have analytical solutions for eigenvectors and eigenenergies.

In Section III, we present analytical solutions and numerical results for the bandgaps, wave functions, and optical matrix elements at the band edges as a function of the crystal orientation for strained GaN bulk semiconductors. The analytical expressions will be useful for understanding the crystal orientation effects on the physical properties of GaN. Also, crystal orientation effects of electronic and optical properties in wurtzite GaN-based QWs with PZ and SP polarization are shown. We then conclude in Sec. IV.

# 2 Theory

## 2.1 Wurtzite $6 \times 6$ Hamiltonians with an Arbitrary Crystal Orientations

In this section, we show the general Hamiltonian for wurtzite crystals with an arbitrary crystal orientation. The Hamiltonian for the valence-band structure has been derived by the $\mathbf{k} \cdot \mathbf{p}$ method.[6, 31] The full Hamiltonian for a (0001)-oriented wurtzite crystal can be written as

$$H(\mathbf{k},\epsilon) = \begin{pmatrix} F & -K^* & -H^* & 0 & 0 & 0 \\ -K & G & H & 0 & 0 & \Delta \\ -H & H^* & \lambda & 0 & \Delta & 0 \\ 0 & 0 & 0 & F & -K & H \\ 0 & 0 & \Delta & -K^* & G & -H^* \\ 0 & \Delta & 0 & H^* & -H & \lambda \end{pmatrix} \begin{matrix} |U_1> \\ |U_2> \\ |U_3> \\ |U_4>, \\ |U_5> \\ |U_6> \end{matrix} \tag{1}$$

where

$$\begin{aligned} F &= \Delta_1 + \Delta_2 + \lambda + \theta, \\ G &= \Delta_1 - \Delta_2 + \lambda + \theta, \\ \lambda &= \frac{\hbar^2}{2m_o}[A_1 {k_z}^2 + A_2({k_x}^2 + {k_y}^2)] + \lambda_\epsilon, \\ \theta &= \frac{\hbar^2}{2m_o}[A_3 {k_z}^2 + A_4({k_x}^2 + {k_y}^2)] + \theta_\epsilon, \\ K &= \frac{\hbar^2}{2m_o}A_5(k_x + ik_y)^2 + D_5\epsilon_+, \\ H &= \frac{\hbar^2}{2m_o}A_6(k_x + ik_y)k_z + D_6\epsilon_{z+}, \\ \lambda_\epsilon &= D_1\epsilon_{zz} + D_2(\epsilon_{xx} + \epsilon_{yy}), \\ \theta_\epsilon &= D_3\epsilon_{zz} + D_4(\epsilon_{xx} + \epsilon_{yy}), \\ \epsilon_+ &= \epsilon_{xx} - \epsilon_{yy} + 2i\epsilon_{xy}, \\ \epsilon_{z+} &= \epsilon_{xz} + i\epsilon_{yz}, \\ \Delta &= \sqrt{2}\Delta_3. \end{aligned} \tag{2}$$

Here, the $A_i$s are the valence-band effective-mass parameters which are similar to the Luttinger parameters in a zincblende crystal, the $D_i$s are the deformation potentials for wurtzite crystals, $k_i$ is the wave vector, $\epsilon_{ij}$ is the strain tensor, $\Delta_1$ is the crystal-field split energy, and $\Delta_2$ and $\Delta_3$ account for spin-orbit interactions. The bases for the Hamiltonian are defined as[6]

$$\begin{aligned} |U_1> &= -\frac{1}{\sqrt{2}}|(X + iY)\uparrow>, \\ |U_2> &= \frac{1}{\sqrt{2}}|(X - iY)\uparrow>, \\ |U_3> &= |Z\uparrow>, \\ |U_4> &= \frac{1}{\sqrt{2}}|(X - iY)\downarrow>, \\ |U_5> &= -\frac{1}{\sqrt{2}}|(X + iY)\downarrow>, \\ |U_6> &= |Z\downarrow>. \end{aligned} \tag{3}$$

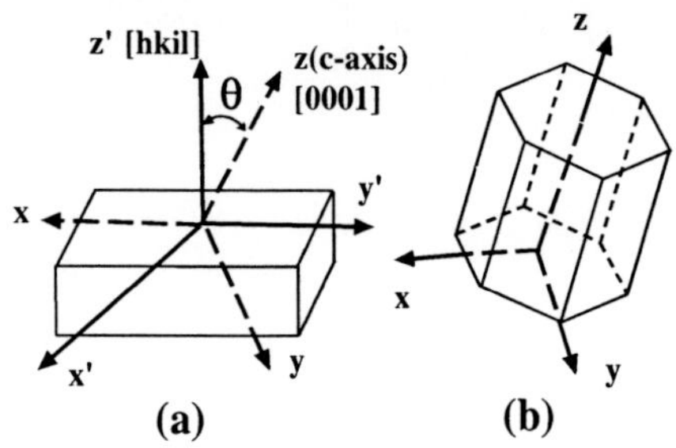


Figure 1: (a) Configuration of the coordinate systems $(x', y', z')$ in $(hkil)$-oriented crystals. The growth axis or the $z'$-axis is normal to the substrate surface $(hkil)$ and the coordinate system $(x, y, z)$ denotes the primary crystallographic axes. (b) The coordinate system in a wurtzite primitive cell. The angle $\theta$ is defined as the angle between the growth direction $\hat{\mathbf{z}}'$ and the c-axis; $\theta = 0^o$ corresponds to $\hat{\mathbf{z}}'$=(0001) growth direction and $\theta = 90^o$ corresponds to $\hat{\mathbf{z}}'$=(10$\bar{1}$0) growth direction. (From Ref. [31])

The Hamiltonian for an arbitrary crystal orientation can be obtained using a rotation matrix

$$U = \begin{pmatrix} \cos\theta\cos\phi & \cos\theta\sin\phi & -\sin\theta \\ -\sin\phi & \cos\phi & 0 \\ \sin\theta\cos\phi & \sin\theta\sin\phi & \cos\theta \end{pmatrix}. \tag{4}$$

The rotation of the Euler angles $\theta$ and $\phi$ transforms physical quantities from the $(x, y, z)$ coordinates to the $(x', y', z')$ coordinates. We consider only the $\theta$ dependence of physical quantities in the following due to the hexagonal symmetry.[9] The $z$-axis corresponds to the $c$-axis [0001] and the growth axis (defined as the $z'$-axis) is normal to the QW plane $(hkil)$, as shown in Fig. 1. The relation between the coordinate systems for vectors and tensors is expressed as[32, 33]

$$\begin{aligned} k'_i &= U_{i\alpha}k_\alpha, \\ \epsilon'_{ij} &= U_{i\alpha}U_{j\beta}\epsilon_{\alpha\beta}, \\ C'_{ijkl} &= U_{i\alpha}U_{j\beta}U_{i\gamma}U_{j\delta}C_{\alpha\beta\gamma\delta}, \end{aligned} \tag{5}$$

where summation over repeated indices is indicated.

The strain coefficients in the $(x, y, z)$ coordinates for a general crystal orientation are determined from the condition that the layer is grown pseudomorphically and these strain coefficients should minimize the strain energy of the layer simultaneously[9, 34]. We define the unit vectors $\hat{\mathbf{x}}'$, $\hat{\mathbf{y}}'$, and $\hat{\mathbf{z}}'$ along the $\mathbf{x}'$, $\mathbf{y}'$, and $\mathbf{z}'$ axes and they are related to unit vectors $\hat{\mathbf{x}}, \hat{\mathbf{y}}$, and $\hat{\mathbf{z}}$ along the $\mathbf{x}, \mathbf{y}$, and $\mathbf{z}$ axes through the rotation matrix (4). The hexagonal primitive translational vectors are

$$\begin{aligned}\boldsymbol{\alpha}_i &= a_i\hat{\mathbf{x}}, \\ \boldsymbol{\beta}_i &= \frac{a_i}{2}\hat{\mathbf{x}} + \frac{\sqrt{3}a_i}{2}\hat{\mathbf{y}}, \\ \boldsymbol{\gamma}_i &= c_i\hat{\mathbf{z}}, \end{aligned} \tag{6}$$

where $a$ is the lattice constant and $i$ labels the epilayer($e$) and the substrate($s$). When the crystal is translated, the primitive translational vectors become[2]

$$\begin{aligned}\boldsymbol{\alpha}_i'' &= a_i\hat{\mathbf{x}}'', \\ \boldsymbol{\beta}_i'' &= \frac{a_i}{2}\hat{\mathbf{x}}'' + \frac{\sqrt{3}a_i}{2}\hat{\mathbf{y}}'', \\ \boldsymbol{\gamma}_i'' &= c_i\hat{\mathbf{z}}'', \end{aligned} \tag{7}$$

where

$$\begin{aligned}\hat{\mathbf{x}}'' &= (1+\epsilon_{xx})\hat{\mathbf{x}} + \epsilon_{xy}\hat{\mathbf{y}} + \epsilon_{xz}\hat{\mathbf{z}}, \\ \hat{\mathbf{y}}'' &= \epsilon_{yx}\hat{\mathbf{x}} + (1+\epsilon_{yy})\hat{\mathbf{y}} + \epsilon_{yz}\hat{\mathbf{z}}, \\ \hat{\mathbf{z}}'' &= \epsilon_{zx}\hat{\mathbf{x}} + \epsilon_{zy}\hat{\mathbf{y}} + (1+\epsilon_{zz})\hat{\mathbf{z}}. \end{aligned} \tag{8}$$

When the first atomic layers are deposited on the substrate, these layers will be strained to match the substrate and a pseudomorphic (or coherent) interface will be formed. Thus, the condition for a pseudomorphic interface means that the projections of the strain-distorted primitive translational vectors of the epilayer and the substrate on the growth plane should be equal:

$$\begin{aligned}\boldsymbol{\alpha}_e'' \cdot \hat{\mathbf{x}}' &= \boldsymbol{\alpha}_s'' \cdot \hat{\mathbf{x}}', \\ \boldsymbol{\alpha}_e'' \cdot \hat{\mathbf{y}}' &= \boldsymbol{\alpha}_s'' \cdot \hat{\mathbf{y}}', \end{aligned} \tag{9}$$

with analogous conditions on $\beta''$ and $\gamma''$. Here, we assume that the substrate is unstrained. Then, the constraints (9) yield the following relations for the strain tensors:

$$\begin{aligned}\epsilon_{xx} &= \epsilon_{xx}^{(0)} + \epsilon_{xz}\frac{\sin\theta}{\cos\theta}, \\ \epsilon_{yy} &= \epsilon_{xx}^{(0)}, \\ \epsilon_{zz} &= \epsilon_{xz}\frac{\cos\theta}{\sin\theta} + \epsilon_{zz}^{(0)}, \\ \epsilon_{xy} &= \epsilon_{yz} = 0, \end{aligned} \tag{10}$$

where $\epsilon_{xx}^{(0)} = (a_s - a_e)/a_e$ and $\epsilon_{zz}^{(0)} = (c_s - c_e)/c_e$. Under the engineering notation described further in Subsection II.E, the strain energy density is given by

$$W = \frac{1}{2}\left[C_{11}\epsilon_{xx}^2 + C_{11}\epsilon_{yy}^2 + C_{33}\epsilon_{zz}^2 + 2C_{12}\epsilon_{xx}\epsilon_{yy} + 2C_{13}(\epsilon_{xx}\epsilon_{zz} + \epsilon_{yy}\epsilon_{zz}) + 4C_{44}\epsilon_{xz}^2\right]. \tag{11}$$

Using the above relations, the strain energy can be expressed through only one strain component $\epsilon_{xz}$, which can be found by minimizing the strain energy with respect to the variable $\epsilon_{xz}$. This procedure gives the following expression for $\epsilon_{xz}$

$$\epsilon_{xz} = -\frac{\left[(c_{11}+c_{12}+c_{13}\epsilon_{zz}^{(0)}/\epsilon_{xx}^{(0)})\sin^2\theta + (2c_{13}+c_{33}\epsilon_{zz}^{(0)}/\epsilon_{xx}^{(0)})\cos^2\theta\right]\epsilon_{xx}^{(0)}\cos\theta\sin\theta}{c_{11}\sin^4\theta + 2(c_{13}+2c_{44})\sin^2\theta\cos^2\theta + c_{33}\cos^4\theta}. \quad (12)$$

Hence, we obtain a $6 \times 6$ Hamiltonian in the $(x', y', z')$ coordinates by substituting the transformation relation for the vector $k$ in Eq. (5) and the strain coefficients for a general crystal orientation into the Hamiltonian (1) for the (0001) orientation.

## 2.2 Conduction Band Edge of Bulk Wurtzite Semiconductors

The conduction bands of a bulk wurtzite crystal can be characterized by a parabolic band model, and the effective-mass Hamiltonian can be written as[6]

$$H(\mathbf{k}) = \frac{\hbar^2}{2}\left(\frac{k_x^2 + k_y^2}{m_e^t} + \frac{k_z^2}{m_e^z}\right) + E_c^0 + P_{c\epsilon}, \quad (13)$$

where the band edge energy is given by

$$E_c^0 = \Delta_1 + \Delta_2 + E_g + P_{c\epsilon}, \quad (14)$$

which includes a hydrostatic energy shift,

$$P_{c\epsilon} = a_{cz}\epsilon_{zz} + a_{ct}(\epsilon_{xx} + \epsilon_{yy}) \quad (15)$$

where $a_{cz}$ and $a_{ct}$ are the conduction-band deformation potentials along the c-axis and perpendicular to the c-axis, respectively. We take $a_{cz} = a_{ct}$ for simplicity. The conduction band wavefunction of a bulk wurtzite semiconductor is of the form

$$\Psi^{c\eta}(\mathbf{r}, \mathbf{k}) = \frac{1}{\sqrt{V}} e^{i\mathbf{k}\cdot\mathbf{r}} |S\eta>, \quad (16)$$

where $\eta = \uparrow$ or $\downarrow$ and $S$ is a spherically symmetric wavefunction.

## 2.3 Valence Band Edges of Bulk Wurtzite Semiconductors

At the band edges, $k'_x = k'_y = k'_z = 0$, the Hamiltonian in the $(x', y', z')$ coordinates can be simplified

$$H'(\mathbf{k}'=0,\epsilon) = \begin{pmatrix} F' & -K'^* & -H'^* & 0 & 0 & 0 \\ -K' & G' & H' & 0 & 0 & \Delta \\ -H' & H'^* & \lambda' & 0 & \Delta & 0 \\ 0 & 0 & 0 & F' & -K' & H' \\ 0 & 0 & \Delta & -K'^* & G' & -H'^* \\ 0 & \Delta & 0 & H'^* & -H' & \lambda' \end{pmatrix} \begin{matrix} |U_1>' \\ |U_2>' \\ |U_3>' \\ |U_4>', \\ |U_5>' \\ |U_6>' \end{matrix} \quad (17)$$

where

$$\begin{aligned}
F' &= \Delta_1 + \Delta_2 + \lambda'_\epsilon + \theta'_\epsilon, \\
G' &= \Delta_1 - \Delta_2 + \lambda'_\epsilon + \theta'_\epsilon, \\
K' &= D_5(\epsilon_{xx} - \epsilon_{yy}), \\
H' &= D_6\epsilon_{xz}, \\
\lambda'_\epsilon &= D_1\epsilon_{zz} + D_2(\epsilon_{xx} + \epsilon_{yy}), \\
\theta'_\epsilon &= D_3\epsilon_{zz} + D_4(\epsilon_{xx} + \epsilon_{yy}), \\
\Delta &= \sqrt{2}\Delta_3,
\end{aligned} \tag{18}$$

and the strain coefficients $\epsilon_{ij}$ for a general crystal orientation $\theta$ are given by Eqs. (10) and (12) in Section II.A. Note that $K'$ and $H'$ are not zero because $\epsilon_{xx} \neq \epsilon_{yy}$ and $\epsilon_{xz} \neq 0$ in the general coordinates. Also, all elements are constant. This Hamiltonian can be block-diagonalized following a similar procedure to that used in Ref. [6]. The transformation matrix is defined as

$$T = \begin{pmatrix} \alpha^* & 0 & 0 & \alpha & 0 & 0 \\ 0 & \beta & 0 & 0 & \beta^* & 0 \\ 0 & 0 & \beta^* & 0 & 0 & \beta \\ \alpha^* & 0 & 0 & -\alpha & 0 & 0 \\ 0 & \beta & 0 & 0 & -\beta^* & 0 \\ 0 & 0 & -\beta^* & 0 & 0 & \beta \end{pmatrix}, \tag{19}$$

where

$$\alpha = \frac{1}{\sqrt{2}}e^{i(3\pi/4)}, \beta = \frac{1}{\sqrt{2}}e^{i(\pi/4)} \tag{20}$$

These expressions are similar to those in Ref. [6] except that $k'_x = k'_y = k'_z = 0$, and $K'$ and $H'$ contain non-zero strain components. The block-diagonalized Hamiltonian, $H'' = T^*H'T^t$, is given by

$$H'' = \begin{pmatrix} H'^U & 0 \\ 0 & H'^L \end{pmatrix}, \tag{21}$$

where the upper three-by-three Hamiltonian is

$$H'^U = \begin{pmatrix} F' & K' & -iH' \\ K' & G' & \Delta - iH' \\ iH' & \Delta + iH' & \lambda' \end{pmatrix} \tag{22}$$

and the lower three-by-three Hamiltonian is

$$H'^L = \begin{pmatrix} F' & K' & iH' \\ K' & G' & \Delta + iH' \\ -iH' & \Delta - iH' & \lambda' \end{pmatrix}. \tag{23}$$

The lower Hamiltonian is the complex conjugate of the upper Hamiltonian. Therefore, they have exactly the same eigenvalues since the eigenenergies are real. However, for the QW structure, there will be a lifting of the Kramers degeneracy[35, 36] in the in-plane dispersion relations because of the piezoelectric field. The piezoelectric field in bulk crystal does not change the in-plane dispersion curves as discussed by Smith *et al.*[34]. The wave functions of the lower Hamiltonian are the complex conjugates of the corresponding wave functions of the upper Hamiltonian. The eigenequation for the valence band energy $E'$ at the band edge is given by

$$\begin{pmatrix} F'-E' & K' & -iH' \\ K' & G'-E' & \Delta - iH' \\ iH' & \Delta + iH' & \lambda' - E' \end{pmatrix} \begin{pmatrix} g'_1 \\ g'_2 \\ g'_3 \end{pmatrix} = 0 \tag{24}$$

The eigenequation is a third-order polynomial equation, which can always be solved analytically. These three analytical solutions can be expressed as[7, 37]

$$\begin{aligned} E'_1 &= 2(S_1+S_2) - \frac{C_2}{3}, \\ E'_2 &= -\frac{1}{2}(S_1+S_2) - \frac{C_2}{3} + i\frac{\sqrt{3}}{2}(S_1-S_2), \\ E'_3 &= -\frac{1}{2}(S_1+S_2) - \frac{C_2}{3} - i\frac{\sqrt{3}}{2}(S_1-S_2), \end{aligned} \tag{25}$$

where

$$\begin{aligned} q &= \frac{1}{3}C_1 - \frac{1}{9}C_2^2, \\ r &= \frac{1}{6}(C_1C_2 - 3C_0) - \frac{C_2^3}{27}, \\ S_1 &= [r + \sqrt{q^3+r^2}]^{1/3}, \\ S_2 &= [r - \sqrt{q^3+r^2}]^{1/3}, \end{aligned} \tag{26}$$

and

$$\begin{aligned} C_0 &= -(F'G'\lambda' + 2K'H'^2 - H'^2G' - K'^2\lambda' - F'\Delta^2 - F'H'^2), \\ C_1 &= G'\lambda' + F'\lambda' + F'G' - \Delta^2 - 2H'^2 - K'^2, \\ C_2 &= -(G' + F' + \lambda). \end{aligned} \tag{27}$$

The envelope functions corresponding to the three eigenvalues are determined by

$$\begin{pmatrix} g'_1 \\ g'_2 \\ g'_3 \end{pmatrix} = \frac{1}{D} \begin{pmatrix} D_1 \\ D_2 \\ D_3 \end{pmatrix} \tag{28}$$

where

$$\begin{aligned} D_1 &= iH'(G'-E')+(\Delta-iH')K', \\ D_2 &= iH'K'(F'-E')(\Delta-iH'), \\ D_3 &= (F'-E')(G'-E')-K'^2, \end{aligned} \tag{29}$$

and

$$D=\sqrt{|D_1|^2+|D_2|^2+|D_3|^2}. \tag{30}$$

The complete Bloch wave functions for the valence band are given by

$$\Psi'^v(\mathbf{r}',\mathbf{k}'=0\,)=\frac{1}{\sqrt{V}}e^{i\mathbf{k}'\cdot\mathbf{r}'}\sum_{i=1}^{3}g'_i(\mathbf{k}')|i>', \tag{31}$$

where

$$\begin{aligned} |1>' &= \alpha^*|U_1>'+\alpha|U_4>', \\ |2>' &= \beta|U_2>'+\beta^*|U_5>', \\ |3>' &= \beta^*|U_3>'+\beta|U_6>', \end{aligned} \tag{32}$$

and we can obtain similar expressions for the lower Hamiltonian (23).

## 2.4 Polarization-Dependent Interband Optical-Matrix Elements of Bulk Wurtzite Semiconductors

The optical momentum matrix elements in the $(x', y', z')$ coordinates are given by

$$|\hat{\mathbf{e}}'\cdot\mathbf{M}'^{\eta}|^2=|<\Psi'^{c\eta}|\hat{\mathbf{e}}'\cdot\mathbf{p}'|\Psi'^v>|^2. \tag{33}$$

Using the expressions for the band-edge wavefunctions of the conduction and valence bands given in Eqs. (16),(28) and (31), we obtain the polarization-dependent interband momentum-matrix elements:
TE-polarization ($\hat{\mathbf{e}}'=\hat{\mathbf{x}}'$ or $\hat{\mathbf{y}}'$ ):

$$|M'^{\eta}_{x'}|^2 = \left|-\frac{1}{\sqrt{2}}g'_1\alpha^*\cos\theta<S|p_x|X>+\frac{1}{\sqrt{2}}g'_2\beta\cos\theta<S|p_x|X>-g'_3\beta^*\sin\theta<S|p_z|Z>\right|^2 \tag{34}$$

for $\sigma=U$

$$=\left|-\frac{1}{\sqrt{2}}g'_4\alpha^*\cos\theta<S|p_x|X>+\frac{1}{\sqrt{2}}g'_5\beta\cos\theta<S|p_x|X>+g'_6\beta^*\sin\theta<S|p_z|Z>\right|^2 \tag{35}$$

for $\sigma=L$

TM-polarization ($\hat{\mathbf{e}}'=\hat{\mathbf{z}}'$ ):

$$|M'^{\eta}_{z'}|^2=\left|-\frac{1}{\sqrt{2}}g'_1\alpha^*\sin\theta<S|p_x|X>+\frac{1}{\sqrt{2}}g'_2\beta\sin\theta<S|p_x|X>+g'_3\beta^*\cos\theta<S|p_z|Z>\right|^2 \tag{36}$$

for $\sigma = U$

$$= \left| -\frac{1}{\sqrt{2}} g'_4 \alpha^* \sin\theta < S|p_x|X > + \frac{1}{\sqrt{2}} g'_5 \beta \sin\theta < S|p_x|X > - g'_6 \beta^* \cos\theta < S|p_z|Z > \right|^2 \quad (37)$$

for $\sigma = L$

The above analytical results include the crystal orientation dependence, and the band-edge matrix elements have been derived [3] using the same procedure as in the derivation of the Kane's parameters ($P_1$ and $P_2$):

$$\begin{aligned} P_x &= P_y =< S|p_x|X >=< S|p_y|Y >= \frac{m_o}{\hbar} P_2, P_z =< S|p_z|Z >= \frac{m_o}{\hbar} P_1, \\ P_1^2 &= \frac{\hbar^2}{2m_o}\left(\frac{m_o}{m_e^z} - 1\right)\frac{(E_g + \Delta_1 + \Delta_2)(E_g + 2\Delta_2) - 2\Delta_3^2}{E_g + 2\Delta_2}, \\ P_2^2 &= \frac{\hbar^2}{2m_o}\left(\frac{m_o}{m_e^t} - 1\right)\frac{E_g[(E_g + \Delta_1 + \Delta_2)(E_g + 2\Delta_2) - 2\Delta_3^2]}{(E_g + \Delta_1 + \Delta_2)(E_g + \Delta_2) - \Delta_3^2} \end{aligned} \quad (38)$$

Substituting $E' = E_1'$, $E_2'$, and $E_3'$, we can obtain moment matrix elements for the inter-band optical transitions between the conduction(C) band and the HH, LH, and CH bands, respectively.

## 2.5 Piezoelectric Field and Spontaneous Polarization in a Quantum Well Structure

If a stress $\tau_{jk}$ is applied to GaN crystals, there is an induced piezoelectric polarization with a magnitude proportional to the applied stress $\tau_{jk}$:

$$P_i = d_{ijk}\tau_{jk}, \quad (39)$$

where $d_{ijk}$ are the piezoelectric moduli or piezoelectric tensor elements. The stress $\tau_{ij}$ is related to the strain by

$$\tau_{ij} = c_{ijkl}\epsilon_{kl}, \quad (40)$$

where $c_{ijkl}$ are the stiffness constants of the wurtzite GaN crystal. Here, it is convenient to replace the tensor notation by the the engineering notation for $d_{ijk}$ and $c_{ijkl}$ using their symmetry properties.[33] That is, the second and third subscripts in $d_{ijk}$ and the first two and the last two subscripts in $c_{ijkl}$ are replaced by a single subscript running from 1 to 6 as follows:

$$\begin{pmatrix} \text{tensor notation} & 11 & 22 & 33 & 23,32 & 31,13 & 12,21 \\ \text{matrix notation} & 1 & 2 & 3 & 4 & 5 & 6 \end{pmatrix} \quad (41)$$

Then, the piezoelectric polarization in the $(x, y, z)$ coordinates for a general crystal orientation is given by

$$\begin{pmatrix} P_x \\ P_y \\ P_z \end{pmatrix} = \begin{pmatrix} 0 & 0 & 0 & 0 & d_{15} & 0 \\ 0 & 0 & 0 & d_{15} & 0 & 0 \\ d_{31} & d_{31} & d_{33} & 0 & 0 & 0 \end{pmatrix} \begin{pmatrix} c_{11} & c_{12} & c_{13} & 0 & 0 & 0 \\ c_{12} & c_{11} & c_{13} & 0 & 0 & 0 \\ c_{13} & c_{13} & c_{33} & 0 & 0 & 0 \\ 0 & 0 & 0 & c_{44} & 0 & 0 \\ 0 & 0 & 0 & 0 & c_{44} & 0 \\ 0 & 0 & 0 & 0 & 0 & c_{66} \end{pmatrix} \begin{pmatrix} \epsilon_{xx} \\ \epsilon_{yy} \\ \epsilon_{zz} \\ 0 \\ 2\epsilon_{xz} \\ 0 \end{pmatrix}, \quad (42)$$

where the strain coefficients in the $(x, y, z)$ coordinates for a general crystal orientation are given in Subsection A. We then obtain the polarization components along the $x$, $y$, and $z$ axes:

$$\begin{aligned} P_x &= 2d_{15}c_{44}\epsilon_{xz}, \\ P_y &= 0, \\ P_z &= [d_{31}(c_{11} + c_{12}) + d_{33}c_{13}](\epsilon_{xx} + \epsilon_{yy}) \\ &+ [2d_{31}c_{13} + d_{33}c_{33}]\epsilon_{zz} \end{aligned} \quad (43)$$

The normal polarization with respect to the growth plane(along the growth direction) can be obtained by taking the scalar product between the polarization vector $\mathbf{P}$ in the $(x, y, z)$ coordinates and the unit vector $\hat{\mathbf{z}}'$ shown in Fig. 1:

$$P_z' = \mathbf{P} \cdot \hat{\mathbf{z}}' = P_x \sin\theta + P_z \cos\theta. \quad (44)$$

Similarly, we can obtain the parallel polarization on the growth plane:

$$\begin{aligned} P_x' &= \mathbf{P} \cdot \hat{\mathbf{x}}' = P_x \cos\theta - P_z \sin\theta, \\ P_y' &= \mathbf{P} \cdot \hat{\mathbf{y}}' = 0 \end{aligned} \quad (45)$$

The strain-induced polarization $P_i$ will lead to an electric field $E_i$ given by

$$D_i' = \epsilon_{ij}^{p\,\prime} E_j' + P_i', \quad (46)$$

where $D_i'$ is the electric displacement and $\epsilon_{ij}^{p\,\prime}$ is the permittivity tensor for the epilayer given by

$$\epsilon' = U\epsilon^p U^{-1} \quad (47)$$

and

$$\epsilon^p = \begin{pmatrix} \epsilon_t & 0 & 0 \\ 0 & \epsilon_t & 0 \\ 0 & 0 & \epsilon_z \end{pmatrix}, \quad (48)$$

where $\epsilon_t$ and $\epsilon_z$ are permittivities[38] perpendicular and along the $c$-axes for the (0001) crystal orientation, respectively. Then, the electric displacement $D'_i$ is given by

$$\begin{aligned} D'_x &= (\epsilon_t \cos^2\theta + \epsilon_z \sin^2\theta)E'_x + (\epsilon_t - \epsilon_z)\sin\theta\cos\theta E'_z + P'_x, \\ D'_y &= \epsilon_t E'_y + P'_y, \\ D'_z &= (\epsilon_t - \epsilon_z)\sin\theta\cos\theta E'_x + (\epsilon_t \sin^2\theta + \epsilon_z \cos^2\theta)E'_z + P'_z. \end{aligned} \tag{49}$$

If there are no external charges and $\epsilon_t \simeq \epsilon_z$, $D'_z$ vanishes and the electric field reduces to

$$E'_z \simeq -\frac{P'_z}{\epsilon_t}. \tag{50}$$

The electric fields in the well and barrier due to the SP and PZ polarization along the (0001)-direction for (0001)-oriented wurtzite structure can be estimated from the periodic boundary condition for a superlattice structure as follows:[9]

$$\begin{aligned} F_{z'} &= \frac{(F^b_{SP} + F^b_{PZ} - F^w_{SP} - F^w_{PZ})L^b}{L^b\epsilon^w + L^w\epsilon^b}, \\ F^b_{z'} &= -\frac{L_w}{L_b}F^w_{z'}, \end{aligned} \tag{51}$$

where superscripts $w$ and $b$ mean the well and barrier, respectively, and $L$ and $\epsilon$ are the layer thickness and the static dielectric constant, repectively. The strain-induced piezoelectric field $F^i_{PZ}$ $(i = w, b)$ is given by[8]

$$F^i_{PZ} = \frac{2d^i_{31}}{\epsilon^i}\left(C^i_{11} + C^i_{12} - \frac{2{C^i_{13}}^2}{C^i_{33}}\right)\epsilon'^{i(3)}_{xx}, \tag{52}$$

where $d_{31}$ is the piezoelectric constants. The parameters for spontaneous polarization were taken from Ref. [10].

The SP polarization $P_{SP}$ along the growth direction in the arbitrary crystal orientation is estimated by the relation $P_{SP} = P^{(0001)}_{SP}\cos\theta$ since the SP polarization in WZ (0001) GaN-based QW structures exists along the (0001) axis.[23] The strain-induced piezoelectric polarization normal with respect to the growth plane (along the growth direction) in an arbitrary crystal orientation can be expressed as

$$P_{PZ} = P_x \sin\theta + P_z\cos\theta. \tag{53}$$

## 2.6 Valence Band Structure of a Quantum Well and Self-Consistent Calculations with the Screening Effect

The constant energy contour of the (0001) orientation is isotropic. On the other hand, the valence band structure of the QW structures with general crystal orientations shows anisotropy in the QW plane unlike the (0001)-oriented structure. Thus, the valence band structure and the momentum matrix element should be calcualted as a function of the angle $\Phi$ in the QW plane. That is, the angle is related to $k_{||}$ by $k'_x = k_{||}\cos\Phi$ and $k'_y = k_{||}\sin\Phi$. Here, $k'_x$ and $k'_y$ are the in-plane wave vectors in the $(x', y', z')$ coordinate system. Please note that, for simplicity, we omit the angle $\Phi$ dependence in the notation of physical functions such as the wavefunction and the momentum matrix element.

For a QW structure of a width $L_w$, the valence band structure is calculated by using $6\times6$ Hamiltonian for an arbitrary crystal orientation, which is obtained by using Eqs.(1) and (5). The quantum well potential due to the band-edge discontinuities of the well and barrier materials follows the model of ref.[39], and it is added with the electro-static potential due to the field from the SP and PZ polarization to the effective mass equation, which the envelope functions $\{g_m^{(\nu)}\}$ satisfy as follows.

$$\sum_{\nu'}\left[H_{\nu\nu'}\left(k'_x, k'_y, -i\frac{\partial}{\partial z'}\right) + (V_\nu(z') + qE'_z z')\delta_{\nu\nu'}\right] g_m^{(\nu)}(k'_x, k'_y, z') = E'_m(k'_x, k'_y) g_m^{(\nu)}(k'_x, k'_y, z'), \quad (54)$$

where $\nu, \nu'$=1,2,3,4,5,6 and $E'_z$ is given in Eq.(50). These coupled differential equations can be solved using a finite-difference method.[39] The material parameters for GaN, InN, and AlN are listed in Table III.

The SC band structures and wave functions are obtained by iteratively solving the Schrödinger equation for electrons, the block-diagonalized Hamiltonian for holes, and Poisson's equation [6, 15]. The total potential profiles for the electrons and the holes are

$$\begin{aligned} V_c(z) &= V_{cw}(z) - |e|\phi(z), \\ V_v(z) &= V_{vw}(z) - |e|\phi(z), \end{aligned} \quad (55)$$

where $V_{cw}(z)$ and $V_{vw}(z)$ are the square-well potentials for the conduction band and the valence band, respectively, and $\phi(z)$ is the screening potential induced by the charged carriers and satisfies the Poisson equation

$$\frac{d}{dz}\left(\epsilon(z)\frac{d}{dz}\right)\phi(z) = -|e|[p(z) - n(z)], \quad (56)$$

where $\epsilon(z)$ is the dielectric constant.

The electron and the hole concentrations, $p(z)$ and $n(z)$, are related to the wave functions of the $n$-th conduction subband and the $m$-th valence subband by

$$n(z) = \frac{kTm_e}{\pi\hbar^2}\sum_n |f_n(z)|^2 \ln\left(1 + e^{[E_{fc} - E_{cn}(0)]/kT}\right) \quad (57)$$

and

$$p(z) = \sum_{\sigma=U,L}\sum_m \int dk_{||}\frac{k_{||}}{2\pi}\sum_\nu |g_{mk_{||}}^{\sigma(\nu)}(z)|^2 \left(\frac{1}{1 + e^{[E_{fv} - E_{vm}(k_{||})]/kT}}\right), \quad (58)$$

where $m_e$ is the effective mass of electrons, $\hbar$ is Planck's constant divided by $2\pi$, $n$ and $m$ are the quantized subband indices for the conduction and the valence bands, $E_{fc}$ and

$E_{fv}$ are the quasi-Fermi levels of the electrons and the holes, respectively, $E_{cn}(0)$ is the quantized energy level of the electrons, $E_{vm}(k_{||})$ is the energy for the $m$-th subband in the valence band, $\sigma$ denotes the upper (U) and the lower (L) blocks of the Hamiltonian, $k_{||}$ is the in-plane wave vector, $\nu$ refers to the new bases for the Hamiltonian [51], and $f_n(z)$ and $g_{mk_{||}}^{\sigma(\nu)}(z)$ are the envelope functions in the conduction and the valence bands, respectively. The potential $\phi(z)$ is obtained by integration [2]:

$$\phi(z) = -\int_{-L/2}^{z} E(z')dz', \tag{59}$$

where

$$E(z) = \int_{-L/2}^{z} \frac{1}{\epsilon(z)}\rho(z')dz'. \tag{60}$$

The procedures for the SC calculations consist of the following steps: (i) Start with the potential profiles $V_c$ and $V_v$ with $\phi^{(0)}(z) = 0$ in Eq. (55); (ii) solve the Schrödinger equation (for electrons) and the block-diagonalized Hamiltonian (for holes) with the potential profiles $\phi^{(n-1)}(z)$ in step (i) to obtain band structures and wave functions; (iii) for a given carrier density, obtain the Fermi-energies by using the band structures and the charge distribution by using the wave functions; (iv) solve Poisson's equation to find $\phi^{(n)}(z)$; (v) check if $\phi^{(n)}(z)$ converges to $\phi^{(n-1)}(z)$. If not, set $\phi^{(n)}(z)= w\phi^{(n)}(z)+ (1-w)\phi^{(n-1)}(z)$, $n = n+1$; then, return to step (ii). If yes, the band structures and the wave functions obtained with $\phi^{(n-1)}(z)$ are solutions. An adjustable parameter $w$ ( $0 < w < 1$) is typically set to 0.5 at low carrier densities. With increasing carrier densities, a smaller value of $w$ is needed for rapid convergence.

## 2.7 Polarization-Dependent Interband Optical-Matrix Elements of a Quantum Well

The optical momentum matrix elements in the $(x', y', z')$ coordinates for a general crystal orientation are given by Eq. (33). The polarization-dependent interband momentum-matrix elements for the QW structure can be written as
TE-polarization ($\hat{\mathbf{e}}' = \cos\phi'\hat{\mathbf{x}}' + \sin\phi'\hat{\mathbf{y}}'$ ):

$$\begin{aligned} |\hat{\mathbf{e}}'\cdot\mathbf{M}'^{\uparrow}|^2 &= \Bigg|\cos\phi'\left\{-\frac{1}{\sqrt{2}}\cos\theta P_x < g'^{(1)}_m|\phi_l> +\frac{1}{\sqrt{2}}\cos\theta P_x < g'^{(2)}_m|\phi_l> - \sin\theta P_z < g'^{(3)}_m|\phi_l>\right\} \\ &+ \sin\phi'\left\{-i\frac{1}{\sqrt{2}}P_x < g'^{(1)}_m|\phi_l> -i\frac{1}{\sqrt{2}}P_x < g'^{(2)}_m|\phi_l>\right\}\Bigg|^2 \end{aligned} \tag{61}$$

$$\begin{aligned} |\hat{\mathbf{e}}'\cdot\mathbf{M}'^{\downarrow}|^2 &= \Bigg|\cos\phi'\left\{\frac{1}{\sqrt{2}}\cos\theta P_x < g'^{(4)}_m|\phi_l> -\frac{1}{\sqrt{2}}\cos\theta P_x < g'^{(5)}_m|\phi_l> - \sin\theta P_z < g'^{(6)}_m|\phi_l>\right\} \\ &+ \sin\phi'\left\{-i\frac{1}{\sqrt{2}}P_x < g'^{(4)}_m|\phi_l> -i\frac{1}{\sqrt{2}}P_x < g'^{(5)}_m|\phi_l>\right\}\Bigg|^2 \end{aligned} \tag{62}$$

TM-polarization ($\hat{\mathbf{e}}' = \hat{\mathbf{z}}'$ ):

$$
\begin{aligned}
|\hat{\mathbf{e}}' \cdot \mathbf{M}'^{\uparrow}|^2 &= \left| -\frac{1}{\sqrt{2}} \sin\theta P_x < g'^{(1)}_m|\phi_l > + \frac{1}{\sqrt{2}} \sin\theta P_x < g'^{(2)}_m|\phi_l > \right. \\
&+ \left. \cos\theta P_z < g'^{(3)}_m|\phi_l > \right|^2
\end{aligned} \tag{63}
$$

$$
\begin{aligned}
|\hat{\mathbf{e}}' \cdot \mathbf{M}'^{\downarrow}|^2 &= \left| \frac{1}{\sqrt{2}} \sin\theta P_x < g'^{(4)}_m|\phi_l > - \frac{1}{\sqrt{2}} \sin\theta P_x < g'^{(5)}_m|\phi_l > \right. \\
&+ \left. \cos\theta P_z < g'^{(6)}_m|\phi_l > \right|^2 ,
\end{aligned} \tag{64}
$$

where $g'^{(\nu)}_m$ ($\nu = 1, 2, 3, 4, 5,$ and 6) is the wave function for the $m$th subband in $(x', y', z')$ coordinates.

## 2.8 Non-Markovian Gain Model with Many-Body Effects

The optical polarization dependent gain spectra are calculated by the non-Markovian gain model with many-body effects under the spontaneous emission transformation method[52, 53]. The optical gain with many-body effects including the effects of anisotropy on the valence band dispersion is given by

$$
g(\omega) = \sqrt{\frac{\mu_o}{\epsilon}} \left(\frac{e^2}{m_o^2\omega}\right) \int_0^{2\pi} d\Phi \int_0^{\infty} dk_{||} \frac{2k_{||}}{(2\pi)^2 L_w} |M_{nm}(k_{||},\Phi)|^2 [f_n^c(k_{||},\Phi) - f_m^v(k_{||},\Phi)] L(\omega, k_{||}, \Phi), \tag{65}
$$

where $\Delta F$ is the quasi-Fermi level separation, $\omega$ is the angular frequency, $\mu_o$ is the vacuum permeability, $\epsilon$ is the dielectric constant, $k_{||}$ and $\Phi$ are the magnitude and angle of the in-plane wave vector in the QW plane, respectively, $L_w$ is the well thickness, $|M_{nm}|^2$ is the momentum matrix element in the strained QW, $f_n^c$ and $f_m^v$ are the Fermi functions for the conduction band states and the valence band states, and $\hbar$ is the Planck constant. The Gaussian line shape function $L(\omega, k_{||}, \Phi)$ renormalized with many-body effects is given by

$$
L(\omega, k_{||}, \Phi) = \frac{(1 - \mathrm{Re}Q(k_{||}, \hbar\omega))\mathrm{Re}\Xi(E_{lm}(k_{||}, \hbar\omega)) - \mathrm{Im}Q(k_{||}, \hbar\omega)\mathrm{Im}\Xi(E_{lm}(k_{||}, \hbar\omega))}{(1 - \mathrm{Re}Q(k_{||}, \hbar\omega))^2 + (\mathrm{Im}Q(k_{||}, \hbar\omega))^2}, \tag{66}
$$

where[52, 53]

$$
\begin{aligned}
\mathrm{Re}\Xi(E_{lm}(k_{||}, \hbar\omega)) &= \sqrt{\frac{\pi\tau_{co}}{2\hbar\Gamma_{cv}(k_{||}, \hbar\omega)}} \\
&\times \exp\left(-\frac{\tau_{co}}{2\hbar\Gamma_{cv}(k_{||}, \hbar\omega)} E_{lm}^2(k_{||}, \hbar\omega)\right)
\end{aligned} \tag{67}
$$

and

$$
\begin{aligned}
\mathrm{Im}\Xi(E_{lm}(k_{||}, \hbar\omega)) &= \frac{\tau_{co}}{\hbar} \int_0^{\infty} \exp\left(-\frac{\Gamma_{cv}(k_{||}, \hbar\omega)\tau_{co}}{2\hbar} t^2\right) \\
&\times \sin\left(\frac{\tau_{co} E_{lm}(k_{||}, \hbar\omega)}{\hbar} t\right) dt.
\end{aligned} \tag{68}
$$

In the above, $E_{lm}(k_{||}, \hbar\omega) = E_l^c(k_{||}) - E_m^v(k_{||}) + E_g + \Delta E_{SX} + \Delta E_{CH} - \hbar\omega$ is the renormalized transition energy between electrons and holes, where $E_g$ is the band-gap of the material. $\Delta E_{SX}$ and $\Delta E_{CH}$ are the screened exchange and Coulomb-hole contributions to the band-gap renormalization. $Q(k_{||}, \hbar\omega)$ is the term related to the excitonic or Coulomb enhancement of the interband transition probability[54]. In the above, $\Gamma_{cv}(k_{||}, \hbar\omega)$ are broadening factors related to the decay rate of the electron and hole wavefunctions. The correlation time $\tau_{co}$ is related to the non-Markovian enhancement of optical gain[52] and is assumed to be constant. The spontaneous polarization $P_{SP}$ along the growth direction in the arbitrary crystal orientation is estimated by the relation $P_{SP} = P_{SP}^{(0001)} \cos\theta$ since the spontaneous polarization in WZ (0001) GaN-based QW structures exists along the (0001) axis[16, 10]. The strain-induced piezoelectric polarization normal with respect to the growth plane (along the growth direction) in an arbitrary crystal orientation can be expressed as[5]

$$P_{PZ} = P_x \sin\theta + P_z \cos\theta, \tag{69}$$

where $P_x = 2d_{15}c_{44}\epsilon_{xz}$, $P_z = [d_{31}(c_{11}+c_{12})+d_{33}c_{13}](\epsilon_{xx}+\epsilon_{yy})+[2d_{31}c_{13}+d_{33}c_{33}]\epsilon_{zz}$, and $c_{ij}$ and $d_{ij}$ are the elastic stiffness and the piezoelectric constants, respectively.

The bandgap renormalization is given as a summation of the Coulomb-hole self energy and the screened-exchange shift.[54] The $\Phi$ dependence of the bandgap renormalization is very small. Here, the $\Phi$ dependence of the bandgap renormalization is neglected for simplicity. The Coulomb-hole contribution to the bandgap renormalization is written as

$$\Delta E_{CH} = -2E_R a_o \lambda_s \ln\left(1 + \sqrt{\frac{32\pi N L_w}{C\lambda_s^3 a_o}}\right), \tag{70}$$

where $N$ is the carrier density and $C$ is a constant usually taken between 1 and 4. The Rydberg constant $E_R$ and the exciton Bohr radius $a_o$ are given by

$$\Delta E_R(eV) = 13.6\frac{m_o/m_r}{(\epsilon/\epsilon_o)^2} \tag{71}$$

and

$$a_o(\mathring{A}) = 0.53\frac{\epsilon/\epsilon_o}{m_o/m_r}, \tag{72}$$

where $m_r$ is the reduced electron-hole mass defined by $1/m_r$=$1/m_e$+$1/m_h$. The exchange contribution to the bandgap renormalization is given by

$$\Delta E_{SX} = -\frac{2E_R a_o}{\lambda_s}\int_0^\infty dk_{||}k_{||}\frac{1+\frac{C\lambda_s a_o k_{||}^2}{32\pi N L_w}}{1+\frac{k_{||}}{\lambda_s}+\frac{C a_o k_{||}^3}{32\pi N L_w}}\left[f_n^c(k_{||}) + 1 - f_m^v(k_{||})\right] \tag{73}$$

The factor $1/(1-Q(k_{||}, \hbar\omega))$ represents the Coulomb enhancement in the Padé approximation. Here, the factor $Q(k_{||}, \hbar\omega)$ is given by[53]

$$Q(k_{||}, \hbar\omega) = i\frac{a_o E_R}{\pi\lambda_s|M_{nm}(k_{||})|}\int_0^\infty dk'_{||}k'_{||}|M_{nm}(k'_{||})|(f_n^c(k'_{||}) - f_m^v(k'_{||}))\Xi(E_{lm}(k'_{||}, \hbar\omega))\Theta(k_{||}, k'_{||}), \tag{74}$$

where

$$\Theta(k_{||}, k'_{||}) = \int_0^{\infty} d\theta \left(1 + \frac{C\lambda_s a_o q_{||}^2}{32\pi N L_w}\right)\left(1 + \frac{q_{||}}{\lambda_s} + \frac{C a_o q_{||}^3}{32\pi N L_w}\right)^{-1} \quad (75)$$

and

$$q_{||} = |\mathbf{k}_{||} - \mathbf{k}_{||}|. \quad (76)$$

# 3 Numerical Results and Discussion

## 3.1 Bulk Semiconductor

In this section, we present numerical results for the crystal orientation dependencies of important physical quantities such as the band gap, strain, wavefunctions, optical matrix elements, and the piezoelectric polarization for a bulk semiconductor.[31] As a numerical example, we consider the situation where an GaN epilayer is grown on an $In_xAl_{1-x}N$ barrier(substrate).

For the following discussions, it should be noted from Fig. 1 that the angle $\theta$ is defined as the angle between the growth direction ($z'$-axis) and and the c-axis (z-axis) as shown in Fig. 1(a). Therefore, $\theta = 0^o$ corresponds to the growth along the c-axis or (0001) orientation and $\theta = 90^o$ corresponds to the growth along the $(10\bar{1}0)$ direction, which is perpendicular to the c-axis. Crystal orientation dependencies of the investigated physical quantities of strained GaN bulk semiconductors are symmetric with respect to $\theta = 90^o$ and will be shown only for the angles between $\theta = 0^o$ and $\theta = 90^o$, except for the piezoelectric field, which will be calculated for $0^o \le \theta \le 180^o$.

### 3.1.1 Band Gap and Strain Tensors

First, we plot the transition energies corresponding to the conduction band (C) to heavy-hole (HH), light-hole (LH), and crystal-field split-off hole(CH) bands. Figs. 2(a) and (b) show these interband transition energies near the band edges as a function of the crystal orientation for (a) -0.5 % compressively (GaN/$In_{0.14}Al_{0.86}N$) and (b) 0.5 % tensile (GaN/$In_{0.22}Al_{0.78}N$) strained GaN crystals. The ground states for holes in the valence band are the HH and LH bands for the compressive and tensile strains, respectively. In the case of (0001) orientation ($\theta = 0^o$) with the compressive strain, it is observed that the energy difference between the HH and LH bands is very small. For comparison, when a zinc-blende layer is grown along the (001) axis with compressive strain, the strain deformation potentials lift the HH band up and reduce its in-plane effective mass.[2] However, for a wurtzite layer grown along the (0001) orientation, the compressive strain shifts both the HH and LH bands by almost the same amount and the in-plane effective masses remain almost the same as those in the unstrained case.[7] Similar results are observed for GaN with a biaxial tensile strain as shown in Fig. 2(b). However, for the compressive strain, the energy difference

Table 1: Physical parameters of GaN.

| parameter | GaN | AlN | InN |
|---|---|---|---|
| Lattice constant ($\AA$) | | | |
| a | 3.1892[40] | 3.112[41] | 3.53[42] |
| c | 5.1850[40] | 4.982[41] | 5.54[42] |
| Energy parameters | | | |
| $E_g$ (eV) | 3.44[43] | 6.28[43] | 1.89[44] |
| $\Delta_{cr}$=$\Delta_1$ (meV) | 22.0[38] | -58.5[45] | 41.0[46] |
| $\Delta_{so}$=3 $\Delta_2$ (meV) | 15.0[38] | 20.4[45] | 1.0[46] |
| $\Delta_3$ = $\Delta_2$ (meV) | | | |
| Conduction band effective masses | | | |
| $m^w_{ez}/m_o$ | 0.20[45] | 0.33[45] | 0.11[47] |
| $m^w_{et}/m_o$ | 0.18[45] | 0.25[45] | 0.11[47] |
| Valence band effective-mass parameters | | | |
| $A_1$ | -6.56[45] | -3.95[45] | -9.09[47] |
| $A_2$ | -0.91[45] | -0.27[45] | -0.63[47] |
| $A_3 = A_2 - A_1$ | | | |
| $A_4 = A_3/2$ | | | |
| $A_5$ | -3.13[45] | -1.95[45] | -4.36[47] |
| $A_6 = (A_3 + 4A_5)/\sqrt{2}$ | | | |
| Deformation potentials (eV) | | | |
| $a_c = -6.4 + a_v$ [42] | -4.60[7] | -4.50[42] | -1.40[7] |
| $D_1$ | -1.70[7] | -2.89[this work] | -1.76[this work] |
| $D_2$ | 6.30[7] | 4.89[this work] | 3.43[this work] |
| $D_3 = D_2 - D_1$ | | | |
| $D_4 = D_3/2$ | | | |
| $D_5$ | -4.00[48] | -3.34[this work] | -2.33[this work] |
| Dielectric constant | | | |
| $\epsilon$ | 10.0[49] | 8.5[49] | 15.3[8] |
| Elastic stiffness constant ($10^{11} dyn/cm^2$) | | | |
| $C_{11}$ | 39.0[50] | 39.8[49] | 27.1[8] |
| $C_{12}$ | 14.5[50] | 14.0[49] | 12.4[8] |
| $C_{13}$ | 10.6[50] | 12.7[49] | 9.4[8] |
| $C_{33}$ | 39.8[50] | 38.2[49] | 20.0[8] |
| $C_{44}$ | 10.5[50] | 9.6[49] | 4.6[42] |
| $C_{66}$ | 12.3[50] | 12.9[49] | 7.4[42] |
| Piezoelectric constant | | | |
| $d_{31}$ ( $\times 10^{-12}$ m/V) | -1.7[8] | -2.0[8] | -1.1[8] |

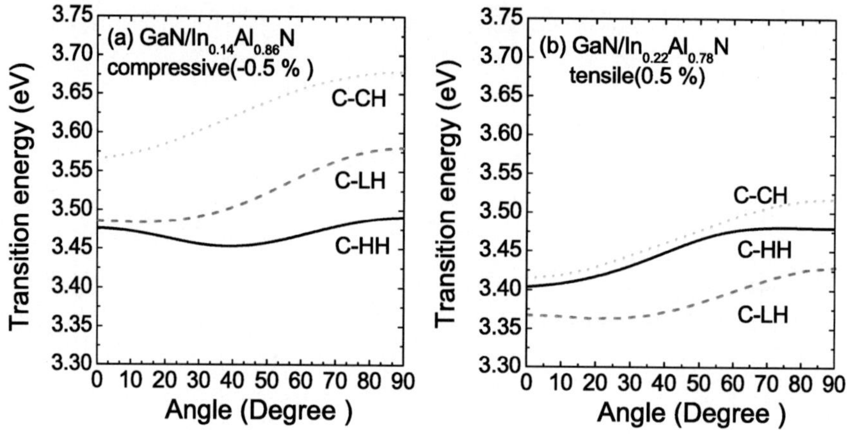


Figure 2: Transition energies at the band-edge as a function of the crystal orientation for (a) -0.5 % compressively($GaN/In_{0.14}Al_{0.86}N$) and (b) 0.5 % tensile($GaN/In_{0.22}Al_{0.78}N$) strained GaN crystals. The angle $\theta$ is defined as the angle between the growth direction $\hat{\mathbf{z}}'$ and the c-axis, as shown in Fig. 1. (From Ref. [31])

between the HH and LH bands increases with the angle $\theta$ between the growth direction and the c-axis and shows a maximum at the $(10\bar{1}0)$ orientation ($\theta = 90^o$). Thus, it is expected that the in-plane effective mass is largely reduced for the $(10\bar{1}0)$ orientation. Similarly, the tensile strain shows that the energy difference between the two lowest bands (LH and HH) increases with the angle $\theta$, until about $60^o$, where a further increase of the angle results in a decrease of the energy difference.

Figures 3(a) and (b) show the band gaps at the band edges as a function of the crystal orientation for various amounts of (a) compressive and (b) tensile strains. The band gaps for compressive and tensile strains correspond to C-HH and C-LH transitions, respectively. The band gap of the compressive strain crystals increases with the strain while that of the tensile strain crystals decreases with the strain. The shift of the band gap as a function of strain is determined mainly by the hydrostatic energy shift of the conduction band edge, the sign of which is positive for the compressive strain and negative for the tensile strain. The band gap as a function of orientation shows a maximum at the $(10\bar{1}0)$ orientation ($\theta = 90^o$) for both compressive and tensile strains.

In Figs. 4(a) and (b), we plot the strain tensor elements as a function of crystal orientation $\theta$ for (a) -0.5 % compressively($GaN/In_{0.14}Al_{0.86}N$) and (b) 0.5 % tensile ($GaN/In_{0.22}Al_{0.78}N$) strained GaN crystals. Note that the component $\epsilon_{yy}$ is constant and the component $\epsilon_{xz}$ is zero for the (0001) orientation ($\theta = 0^o$). The strain components $\epsilon_{xx}$ and $\epsilon_{zz}$ show a maximum at $\theta = 90^o$. Since the crystal orientation dependence of the bands is determined mainly by these stain components, the band gap shows a maximum for the

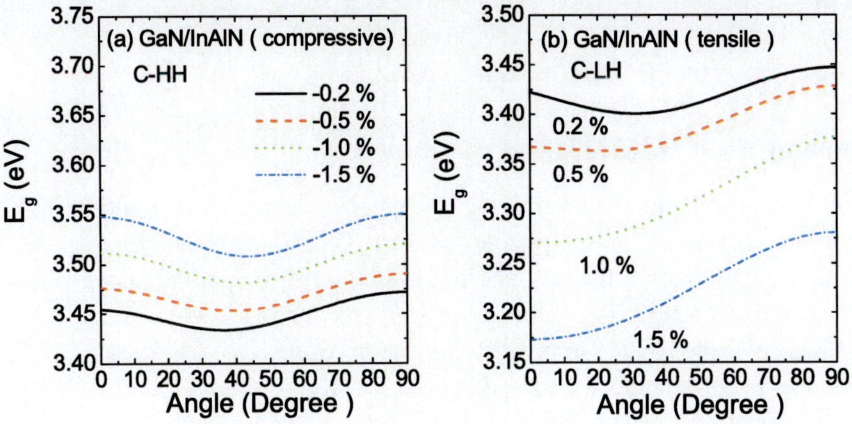


Figure 3: Band gaps near the band edges as a function of the crystal orientation for various (a) compressively and (b) tensile strained GaN crystals. The band gaps for compressive and tensile strains correspond to C-HH and C-LH transitions, respectively. (From Ref. [31])

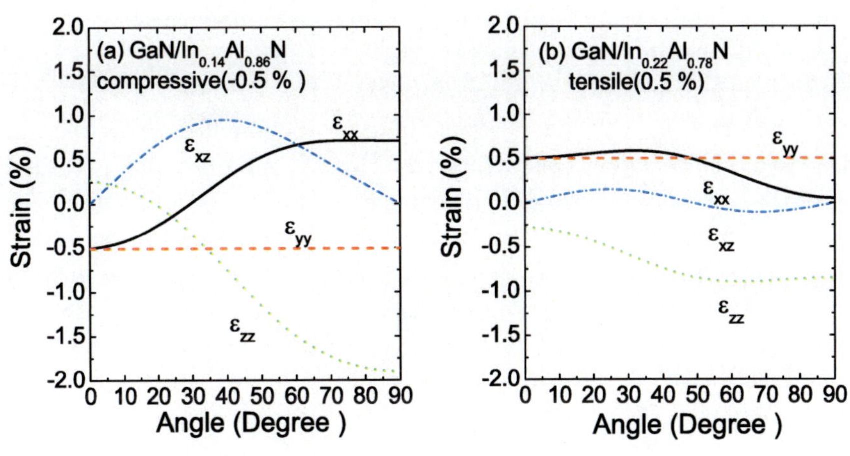


Figure 4: Strain tensor elements as a function of crystal orientation $\theta$ for (a) -0.5 % compressively($GaN/In_{0.14}Al_{0.86}N$) and (b) 0.5 % tensile($GaN/In_{0.22}Al_{0.78}N$) strained GaN crystals. (From Ref. [31])

$(10\bar{1}0)$ growth direction as shown in Fig. 2. In the case of the tensile strain, it is observed that its band gap difference between the $\theta = 0^o$ and $\theta = 90^o$ growth directions is larger than that of the compressive strain. This is attributed to the fact that, for the tensile strain, the band gap is given by the C-LH transition and the orientation dependence of the LH band is larger than that of the HH band as shown in Fig. 2.

### 3.1.2 Energy Band Structures and Wavefunctions

In addition to the band gap, the valence band structures and the wave functions as a function of the crystal orientation show interesting features. The valence band mixing effects can be understood from the energy dispersion relations and their corresponding wave functions of the HH, LH, and CH bands. The valence band structures and the wave functions of the bulk wurtzite GaN with the (0001) orientation has been extensively investigated.[7] However, little work has been done so far to understand the cystal orientation effects of the wave functions for the wurtzite semiconductors.

Figures 5(a)-(c) show the calculated valence band dispersion relations of a -0.5 % compressively strained GaN crystal for (a) $\theta = 0^o$, (b) $\theta = 45^o$, and (c) $\theta = 90^o$. The dispersion relations are calculated numerically from the $6 \times 6$ Hamiltonian in the $(x', y', z')$ coordinates. The full $6 \times 6$ Hamiltonian has to be used since the previous block diagonalization methods are restricted either to the (0001) growth direction or to the zone center ($k'_x = k'_y = k'_z = 0$) case [Eqs.(17)-(23)]. The band edge effective masses can be estimated from a parabolic band fitting to the calculated dispersion relations near the band edges. The crystal orientation effects of the band edge effective mass is useful for the understanding of optical processes such as the optical gain as well as electronic transport processes.

We know from Figs. 5(a)-(c) that the HH band edge effective masses along the $k'_y$ direction are nearly constant independent of the crystal orientation. On the other hand, it is shown that the HH band edge effective masses along the $k'_x$ direction depend on the crystal orientation. The increase of the angle results in the reduction of the effective mass of the HH band along the $k'_x$ direction. For example, the effective mass near the band edge is about $1.1m_o$ and $0.16m_o$ for $\theta = 0^o$ and $90^o$, respectively, along the $k'_x$ direction. This is mainly attributed to the fact that the energy difference between the HH and LH bands increases with the angle between the growth direction and the c-axis from Fig. 2. A similar situation is observed for the LH band of the tensile strain case. In Figs. 6(a)-(c), we plot the calculated valence band dispersion relations of a 0.5 % tensile strained GaN crystal for (a) $\theta = 0^o$, (b) $\theta = 45^o$, and (c) $\theta = 90^o$. For the tensile strain, the top valence band is the LH band and its effective mass along the $k'_x$ direction is reduced with an increasing growth angle $\theta$. Thus, it is expected that the $(10\bar{1}0)$ orientation ($\theta = 90^o$) should improve the performance of both compressively and tensile strained GaN-based lasers due to its reduced effective mass.

Figures 7(a)-(c) show the normalized valence wave functions at the band edge of a -0.5 % compressively strained GaN crystal for the (a) HH, (b) LH, and (c) CH bands. The $v$-th valence band wave function as a function of the crystal orientation is determined by the eigenfunctions of Eq. (28). The mixing ratio of the coefficients ${g'}_1$, ${g'}_2$, and ${g'}_3$ determines the nature of each band. For the (0001) orientation($\theta = 0^o$), the HH band is decoupled from

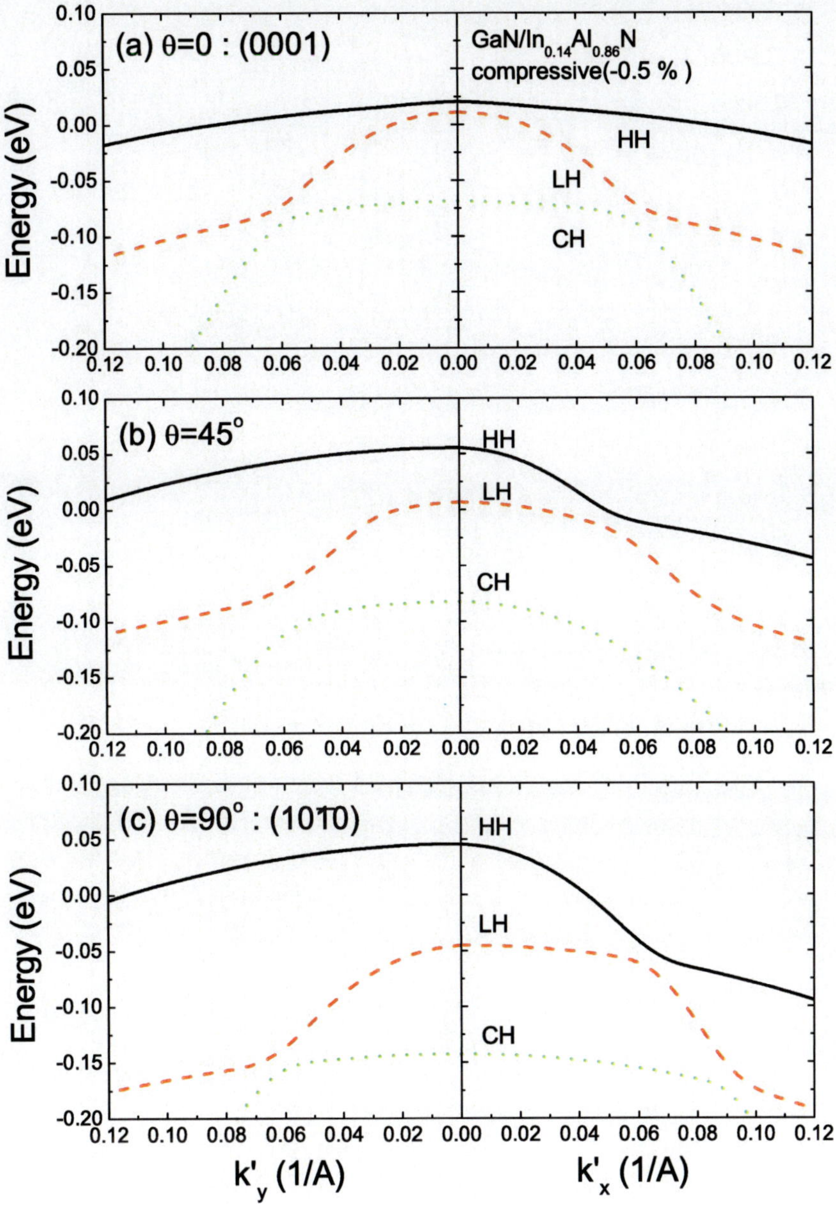


Figure 5: Calculated valence band dispersion relations of a -0.5 % compressively strained GaN crystal for (a) $\theta = 0^o$, (b) $\theta = 45^o$, and (c) $\theta = 90^o$. (From Ref. [31])

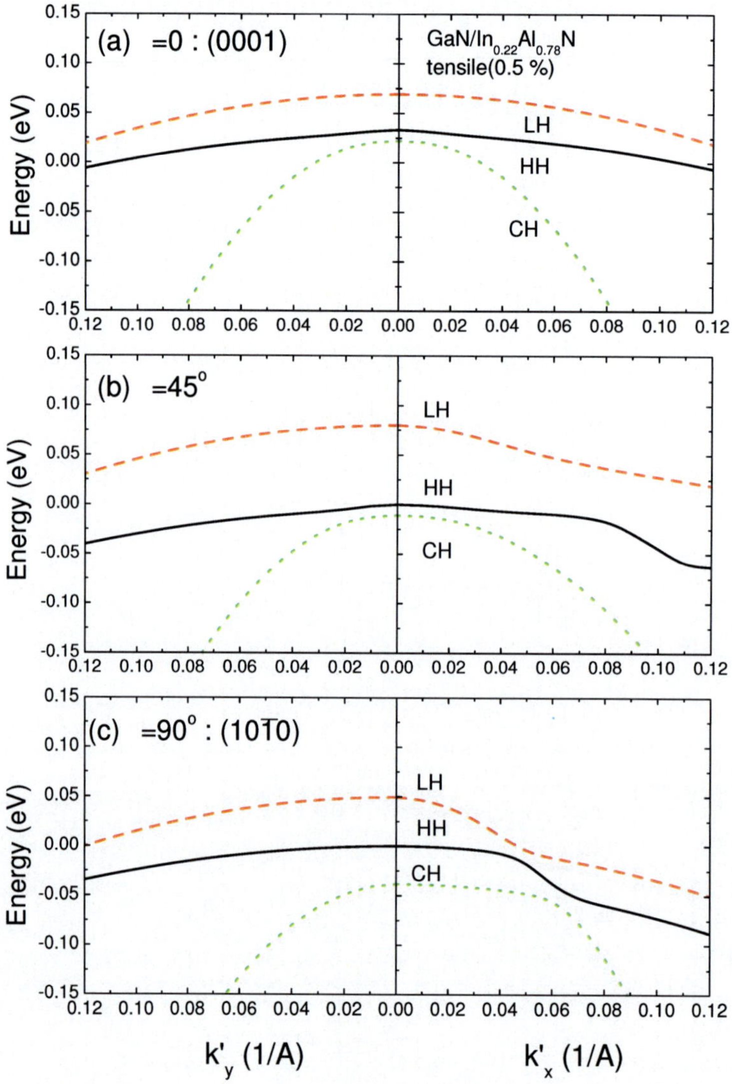


Figure 6: Calculated valence band dispersion relations of a 0.5 % tensile strained GaN crystal for (a) $\theta = 0^o$, (b) $\theta = 45^o$, and (c) $\theta = 90^o$. (From Ref. [31])

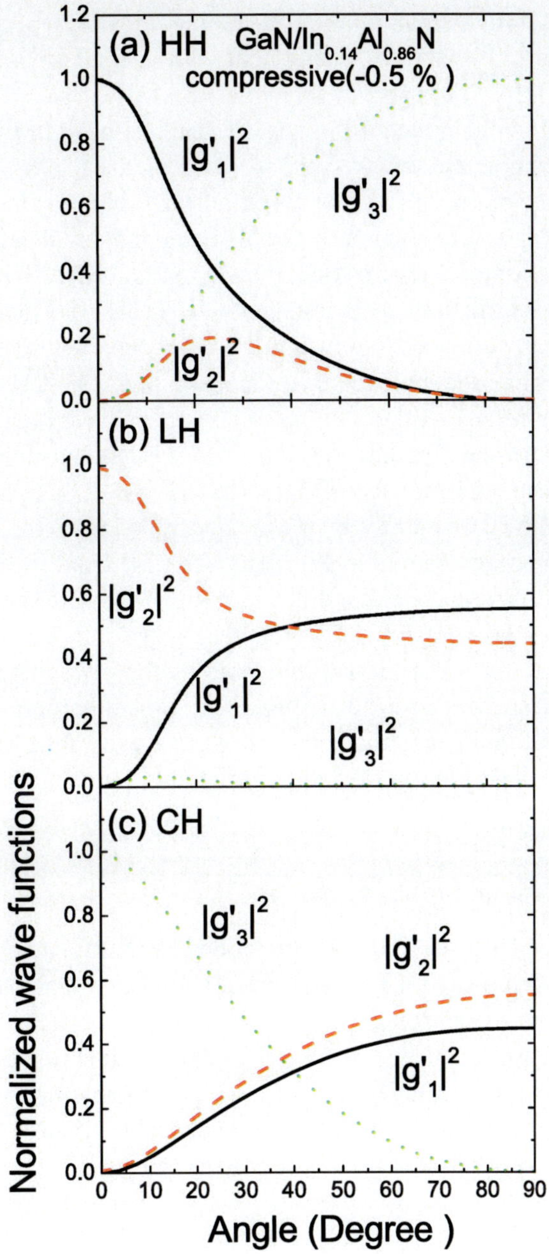


Figure 7: Normalized valence wave functions at the band edge of a -0.5 % compressively strained GaN crystal for the (a) heavy-hole (HH), (b) light-hole (LH,) and (c) crystal-field split-off hole (CH) bands. (From Ref. [31])

the LH and CH bands. Thus, we obtain $g'_1 = 1$ and $g'_2$=$g'_3$=0 as shown in Fig. 7(a) and $g'_1$=0 for the LH and CH bands as shown in Fig. 7(b) and (c). The LH band is dominated by the $g'_2$ component with a small amount of mixing with the $g'_3$ component. Similarly, the CH band is dominated by the $g'_3$ component with a small amount of mixing with the $g'_2$ component.

On the other hand, with an increasing angle $\theta$, the HH band in Fig. 7(a) shows a rapid decrease of the $g'_1$ component and a rapid increase of the $g'_3$ with a small increase of $g'_2$ components. Thus, strong mixing between all three components is observed around $\theta = 25^o$ and the HH band is dominated by the $g'_3$ component with a further increase of the angle. The $g'_2$ and $g'_1$ components for the LH band in Fig. 7(b) also show behavior similar to the HH band. However, the two components are nearly constant in the range above $45^o$. Thus, strong mixing between the $g'_1$ and $g'_2$ components is observed near the $(10\bar{1}0)$ orientation($\theta = 90^o$) while the $g'_3$ component is negligible. The CH band in Fig. 7(c) also shows a strong mixing between the $g'_1$ and $g'_2$ components near the $(10\bar{1}0)$ orientation.

Figures 8(a)-(c) show the normalized valence wave functions at the band edges of a 0.5 % tensile strained GaN crystal for the (a) HH, (b) LH, and (c) CH bands. For the (0001) orientation($\theta = 0^o$), the HH band in Fig. 8(a) is decoupled from the LH and CH bands as in the case of the compressive strain in Fig. 7(a). The LH band in Fig. 8(b) is dominated by the $g'_3$ component while the CH band in Fig. 8(b) is dominated by the $g'_2$ component. With increasing angle, the HH and CH bands generally show a mixing between $g'_1$ and $g'_2$. On the other hand, the LH band is dominated by the $g'_3$ component and the $g'_1$ and $g'_2$ components are nearly zero for all orientations. Since the optical momentum matrix elements are mainly determined by the behavior of these wave functions, it is expected that momentum matrix elements are significantly affected by the crystal orientation as discussed below.

### 3.1.3 Optical Momentum Matrix Elements

From the band edge eigenvalues and wave functions, we calculate the optical momentum matrix elements as functions of the crystal orientation for a GaN semiconductor, which is important for the interband optical transition. Figs. 9(a) and (b) show the interband optical momentum matrix elements at the band edges as functions of crystal orientation $\theta$ for the (a) TE and (b) TM polarizations of a -0.5 % compressive strained GaN crystal. For the (0001) orientation ($\theta = 0^o$), the interband optical matrix elements for the HH and LH bands to the conduction band are dominated by the TE polarization as shown in Fig. 9(a) and the interband transition between the CH band and the conduction band is dominated by the TM polarization as shown in Fig. 9(b). With increasing angle $\theta$, the TE optical matrix element for the C-LH transition rapidly decreases and is nearly zero above $\theta = 30^o$. On the other hand, the TE optical matrix element for the C-HH transition rapidly increases and approaches a constant value above $\theta = 30^o$. The constant value is about two times as large as that of the (0001) orientation. This can be explained by the crystal orientation dependence of the wave functions. As the angle increases, the HH wave function is dominated by the $g'_3$ component as shown in Fig. 7(a), which determines the optical

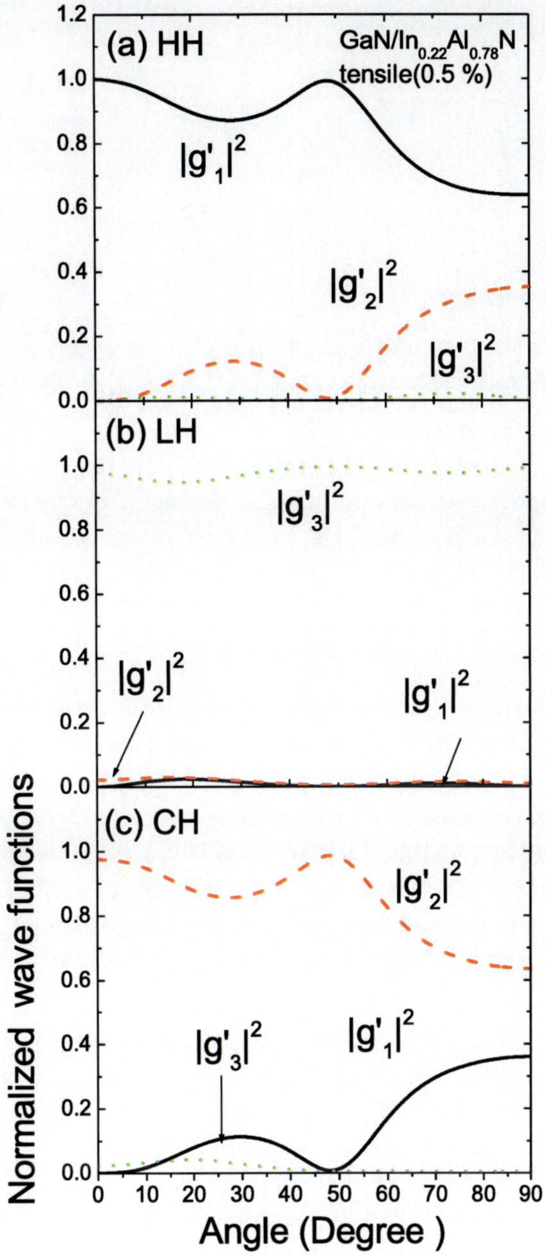


Figure 8: Normalized valence wave functions at the band edges of 0.5 % tensile strained GaN crystal for the (a) HH, (b) LH, and (c) CH bands. (From Ref. [31])

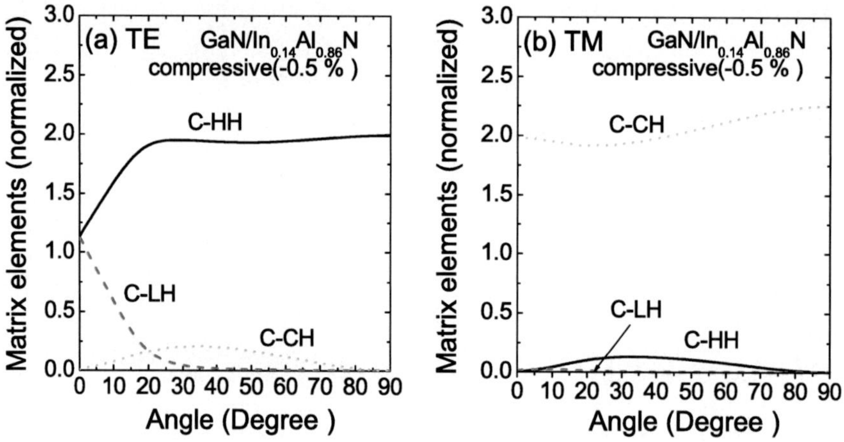


Figure 9: Interband optical momentum matrix elements at the band edges as functions of crystal orientation for the (a) TE and (b) TM polarizations of a -0.5 % compressive strained GaN crystal. (From Ref. [31])

matrix element for the C-HH transition. On the other hand, the LH and CH bands show mixing of the $g'_1$ and $g'_2$ components. However, the $g'_1$ and $g'_2$ contributions to the optical matrix elements of the C-LH and C-CH transitions cancel with each other by an opposite sign as shown in Eq. (61). Similarly, the contributions of the C-HH and C-LH transitions for the TM polarization cancel with each other. Thus, the TM polarization matrix element is dominated by the interband transition between the CH band and the conduction band, as shown in Fig. 9(b).

Figures 10(a) and (b) show the optical momentum matrix elements at the band edges as functions of crystal orientation for the (a) TE and (b) TM polarizations of 0.5 % tensile strained GaN crystals. For the (0001) orientation ($\theta = 0^o$), the interband optical matrix elements for the C-HH and C-CH transitions are dominated by the TE polarization and the C-LH transition is dominated by the TM polarization. On the other hand, at an angle near the ($10\bar{1}0$) orientation($\theta = 90^o$), the interband optical matrix elements for the C-LH and C-HH transitions are dominated by the TE and TM polarizations, respectively. The optical matrix elements for the C-LH transition behave like $\cos^2\theta$ and $\sin^2\theta$ for the TE and TM polarization, respectively, because the wave function for the LH band is dominated by the $g'_3$ component. On the other hand, the optical matrix elements for the C-HH and C-CH transitions are mainly determined by $g'_1$ and $g'_2$ components. They show interesting oscillatory behavior due to the oscillatory behavior of the $g'_1$ and $g'_2$ as shown in Fig. 8.

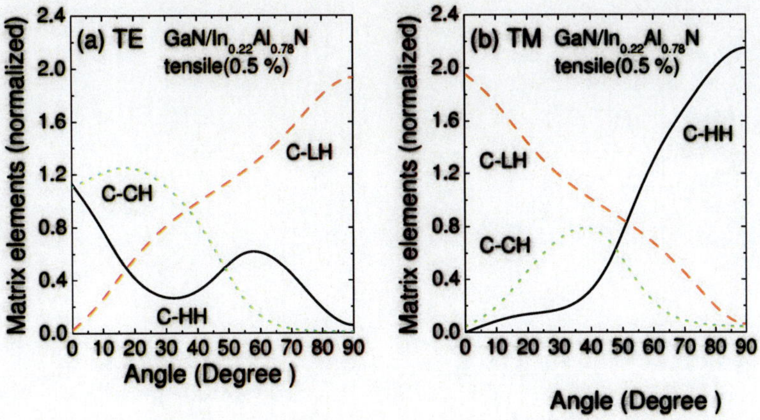


Figure 10: Optical momentum matrix elements at the band edges as functions of crystal orientation for the (a) TE and (b) TM polarizations of a 0.5 % tensile strained GaN crystal. (From Ref. [31])

## 3.2 Quantum Well Structures

### 3.2.1 GaN/AlGaN QW Structures [23, 26, 27, 29]

Figure 11 shows (a) the strain-induced normal PZ polarization and SP polarization with respect to the growth plane for the GaN well and the $Al_{0.2}Ga_{0.8}N$ barrier and (b) the interband transition energies near the band edges for a GaN well as a function of the angle $\theta$ between the growth direction and the c-axis. We consider a GaN/AlGaN MQW structure with the AlGaN barrier under tensile strain and the GaN well lattice-matched to the substrate (sapphire). The normal PZ polarization is important because the QW structures have quantized energy levels along the growth direction. The normal PZ polarization leads to the accumulation or depletion of carriers at the interfaces and creates a piezoelectric field. On which side of the interfaces the accumulation region is produced depends on the type of atomic plane at the interface (Ga or N) in structures such as QWs and semiconductor-insulator-semiconductors.[9]

The strain-induced normal PZ polarization $P^{w}_{PZ}$ in the GaN well is zero because the well is lattice-matched to the substrate. On the other hand, it is shown that the PZ polarization gradually decreases with increasing crystal angle and is nearly zero for crystal angles above about $\theta$=60$^o$. The SP polarization also has a similar crystal orientation dependence. We know that the SP polarization is much larger than the PZ polarization in most of the crystal angle range. The SP and strain-induced PZ polarization becomes zero for the $(10\bar{1}0)$ crystal orientation ($\theta = \pi/2$). Thus, we use the FB model for the $(10\bar{1}0)$-oriented structure. The interband transition energy corresponds to the energy difference between the first conduction subband (C1) and the first valence subband. The ground state for holes in the valence band is the HH band for the (0001)-oriented QW structure. The FB model shows that the interband transition energy slowly decreases with increasing crystal angle. On the

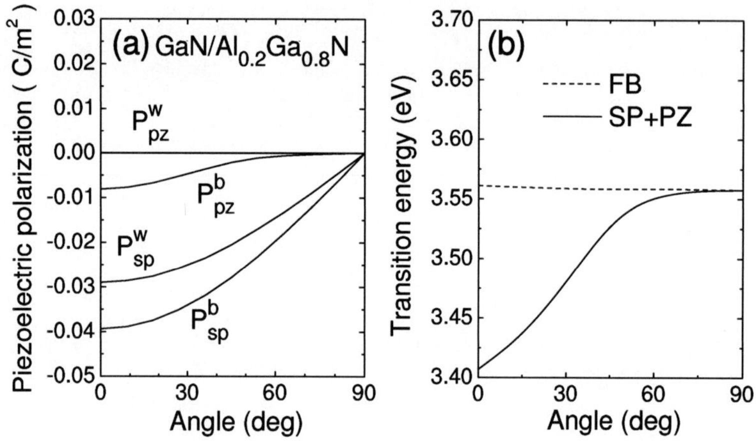


Figure 11: (a) Strain-induced normal PZ polarization and SP polarization with respect to the growth plane for the GaN well and the $Al_{0.2}Ga_{0.8}N$ barrier and (b) interband transition energies near band edges for a GaN well as a function of the angle $\theta$ between the growth direction and the c-axis. (From Ref. [23])

other hand, the interband transition energy for the SC model is significantly redshifted for crystal angles near $\theta$=0$^o$. Also, it is shown that the interband transition energy rapidly increases with the crystal angle due to the reduced PZ and SP polarization effects. These results suggest that an understanding of the role of the PZ and SP polarization as a function of crystal orientation is important to obtain guidelines for sample growth and device design issues.

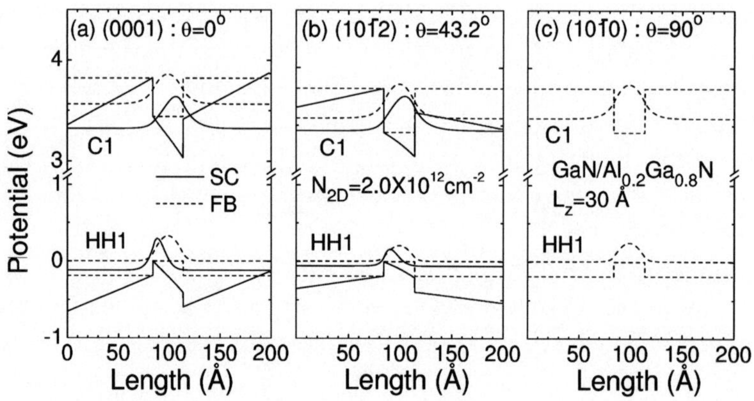


Figure 12: Potential profiles and wave functions (C1 and HH1) at the zone center for the SC model with the SP and PZ polarization of (a) (0001)-, (b) (10$\bar{1}$2)-, and (c) (10$\bar{1}$0)-oriented 30 Å $GaN/Al_{0.2}Ga_{0.8}N$ QW structures. (From Ref. [23])

Figure 12 shows the potential profiles and the wave functions (C1 and HH1) at the zone center for the SC model with the PZ and SP polarization of (a) (0001)-, (b) (10$\bar{1}$2)-, and (c)

(10$\bar{1}$0)-oriented 30 $\AA$ GaN/Al$_{0.2}$Ga$_{0.8}$N QW structures. Here, the barrier width is set to be 84 $\AA$. The SC solutions are obtained at the sheet carrier density of $N_{2D} = 2 \times 10^{12} cm^{-2}$ and the dashed line corresponds to the potential profiles for the FB model without the SP and PZ polarization. The AlGaN barrier is under 0.484% tensile strain and we assume that the layers have a Ga(A1) face. For the Ga face, the SP polarization is directed towards the substrate because the SP polarization for GaN and AlN is found to be negative.[11] Thus, the alignment of the PZ and SP polarization is parallel in the case of tensile strain. If the barrier is under compressive strain, the alignment of the PZ and SP polarization becomes antiparallel.

The potential profiles show that there exists a large internal field in the well due to the difference in the SP polarization between the well and the barrier, although the PZ field in the QW is zero. The estimated field in the well is -1.55 MV/cm for a (0001)-oriented GaN/AlGaN QW with $L_w = 30$ $\AA$. The FB model shows symmetric electron and hole wave functions with respect to the well center while the SC model shows asymmetric electron and hole wave functions. In particular, a (0001)-oriented QW structure shows large spatial separation of the electron and hole wave functions. Also, the ground-state transition energy C1-HH1 is redshifted compared to the FB model due to the quantum-confined Stark effect. On the other hand, it is observed that the spatial separation of wavefunctions is significantly reduced with the increasing crystal orientation. This is due to the decrease of the PZ and SP polarization, which is given as a function of crystal orientation, as shown in Fig. 1(a). Thus, we expect that important properties such as the effective mass and the optical moment matrix elements are affected by the crystal orientation.

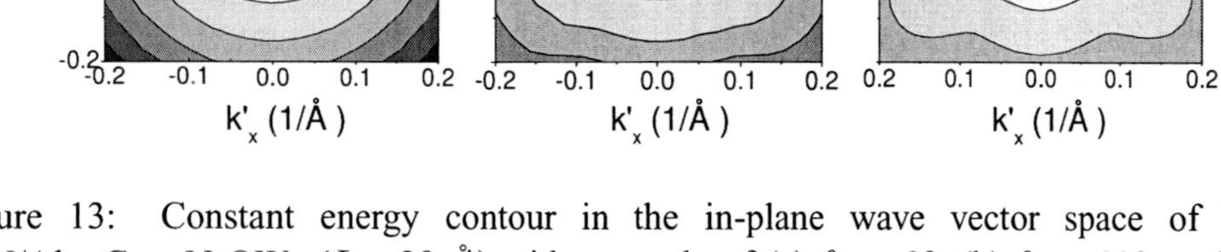


Figure 13: Constant energy contour in the in-plane wave vector space of WZ GaN/Al$_{0.2}$Ga$_{0.8}$N QWs ($L_w$=30 $\AA$) with an angle of (a) $\theta = 0^o$, (b) $\theta = 60^o$, and (d) $\theta = 90^o$. Here, $k'_x$ and $k'_y$ are the in-plane wave vector in the $(x', y', z')$ coordinate system. (From Ref. [26])

Figure 13 shows constant energy contours in the in-plane wave vector space of wurtzite GaN/Al$_{0.2}$Ga$_{0.8}$N QWs ($L_w$=30 $\AA$) with an angle of (a) $\theta = 0^o$, (b) $\theta = 60^o$, and (d) $\theta = 90^o$. Here, $k'_x$ and $k'_y$ are the in-plane wave vectors in the $(x', y', z')$ coordinate system. They are related to $k_{||}$ and $\Phi$ in eq.(65) by $k'_x = k_{||} \cos\Phi$ and $k'_y = k_{||} \sin\Phi$. The angles $\theta = 0^o$ and $\theta = 90^o$ correspond to (0001) and (10$\bar{1}$0) crystal orientations, respectively. The constant energy contour of the (0001) orientation is isotropic. On the other hand, the valence band structure of the GaN/AlGaN QW with general crystal orientations shows anisotropy in the QW plane unlike the (0001)-oriented structure. That is, the constant energy contour of QW structures with $\theta = 60^o$ and $90^o$ shows twofold symmetry. It is observed that the effective mass along the $k'_y$ direction around the topmost valence band is largely reduced for the $(10\bar{1}0)$-oriented structure. This is mainly attributed to the fact that the energy difference between the first and second subbands increases with the angle between the growth and the c-axis.

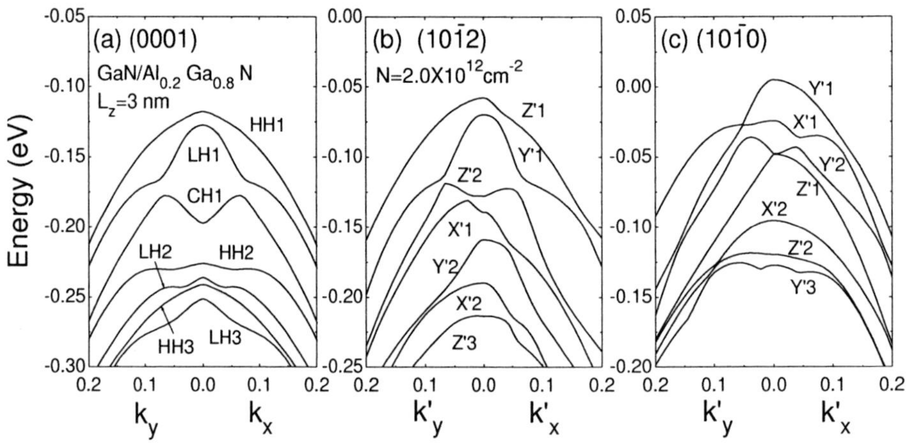


Figure 14: Valence band structures for the SC model with the SP and PZ polarization of (a) (0001)-, (b) $(10\bar{1}2)$-, and (c) $(10\bar{1}0)$-oriented 30 $\AA$ GaN/$Al_{0.2}Ga_{0.8}N$ QW structures. (From Ref. [23])

Figure 14 shows the valence band structures for the SC model with the PZ and SP polarization of (a) (0001)-, (b) $(10\bar{1}2)$-, and (c) $(10\bar{1}0)$-oriented 30 $\AA$ GaN/$Al_{0.2}Ga_{0.8}N$ QW structures. Here, the naming of subbands for $(10\bar{1}2)$- and $(10\bar{1}0)$-oriented structures follows the dominant composition of the wave function at the $\Gamma$ point in terms of $|X' >$, $|Y' >$, and $|Z' >$ bases. The components $P_m^{i(=X',Y',Z')}$ of each wave function are given by

$$
\begin{aligned}
P_m^{X'} &= \left\langle g'^{(3)}_m | g'^{(3)}_m \right\rangle + \left\langle g'^{(6)}_m | g'^{(6)}_m \right\rangle, \\
P_m^{Y'} &= \Big( < g'^{(1)}_m + g'^{(2)}_m | g'^{(1)}_m + g'^{(2)}_m > \\
&+ < g'^{(4)}_m + g'^{(5)}_m | g'^{(4)}_m + g'^{(5)}_m > \Big) /2,
\end{aligned}
$$

and

$$
\begin{aligned}
P_m^{Z'} &= \Big( < g'^{(2)}_m - g'^{(1)}_m | g'^{(2)}_m - g'^{(1)}_m > \\
&+ < g'^{(4)}_m - g'^{(5)}_m | g'^{(4)}_m - g'^{(5)}_m > \Big) /2, \qquad (77)
\end{aligned}
$$

where $g'^{(\nu)}_m$ ($\nu = 1, 2, 3, 4, 5,$ and 6) is the wave function for the $m$th subband in the $(x', y', z')$ coordinates. We know that subband energies lie closer to each other than those in the more widely studied GaAs- and InP-based QW lasers. For example, the subband energy spacing (~10 meV) between the first two subbands observed in Fig. 14(a) is very small compared to that (~70 meV) of InP-based QWs. This is mainly due to the HH effective masses of GaN-based QWs. The valence band structure of $(10\bar{1}2)$- and $(10\bar{1}0)$-oriented structures shows anisotropy in the QW plane, unlike the (0001)-oriented structure. Also, it is observed that the effective masses of the first subbands gradually decrease with the increasing angle between the growth and the c-axis. In particular, the effective mass along $k'_y$ around the topmost valence band is significantly reduced for the $(10\bar{1}0)$-oriented structure. This is mainly attributed to the fact that the energy difference between the first and second subbands increases with the angle between the growth and the c-axis.

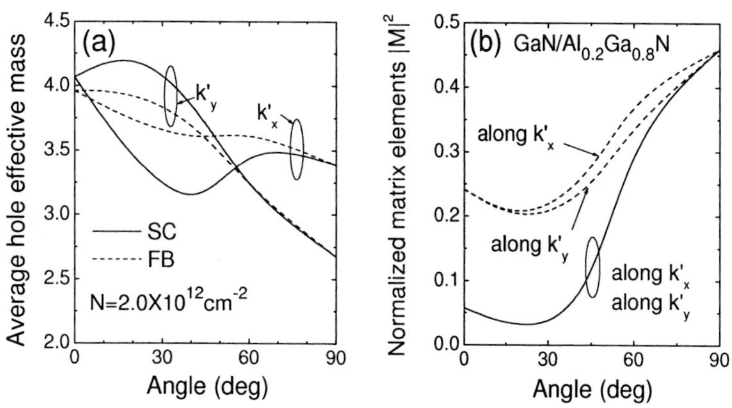


Figure 15: Average hole effective masses and normalized $y'$-polarized optical matrix elements as functions of crystal orientation for 30 Å GaN/Al$_{0.2}$Ga$_{0.8}$N QW structures with the SP and PZ polarization. (From Ref. [23])

Figure 15 shows average hole effective masses and normalized $y'$-polarized optical matrix elements as functions of crystal orientation for 30 Å GaN/Al$_{0.2}$Ga$_{0.8}$N QW structures with the PZ and SP polarization. For comparison, the results for the FB model in the absence of the PZ and SP polarization are also plotted. The average hole effective mass and optical matrix elements of the SC model are calculated at a sheet carrier density of $N_{2D} = 2 \times 10^{12} cm^{-2}$. To estimate the magnitude of the hole effective mass, we consider the parabolic band fitted to the lowest subband of the exact band structure. The effective mass is determined so that, for a given carrier density and the quasi-Fermi level for holes, the carrier density and the quasi-Fermi level agree with those of the exact band structure. Hence, the effective mass of the fitted parabolic band reflects an averaged density of states.[56]

The FB model shows that the average hole effective mass along $k'_x$ gradually decreases with the crystal angle. On the other hand, the SC model shows that the average hole effective mass along $k'_x$ is significantly reduced near the $(10\bar{1}2)$ crystal orientation ($\theta = 43.2^o$). In the case of the average hole effective mass along $k'_y$, both SC and FB models show that their effective mass markedly decreases near the $(10\bar{1}0)$ crystal orientaion. On the basis of these results, we know that the average hole effective mass in the non-(0001) QWs above $\theta = 30^o$, particularly in the $(10\bar{1}0)$ QW, is much smaller than that in the conventional (0001)-oriented QW. The FB model shows that the matrix elements increase with the crystal orientation although they reach a minimum near $\theta = 20^o$. In particular, the $y'$-polarized optical matrix elements of the $(10\bar{1}0)$-oriented QW structure are about twofold bigger than those of the (0001)-oriented QW structure. This can be explained by the fact that, in the case of the $(10\bar{1}0)$-oriented QW structure, the states constituting the topmost valence subband near the band edge are predominantly $|Y' >$-like. This results in significantly enhanced $y'$-polarized optical matrix elements for the C1-$Y'$1 transition in the $(10\bar{1}0)$-oriented QW structure in comparison with the $y$-polarized optical matrix elements in the (0001)-oriented QW structure where $|X >$ and $|Y >$ are equally mixed. In addition, the SC model shows that the matrix elements near the (0001) crystal orientation can be significantly reduced compared to the FB model. For example, the $y$-polarized optical matrix elements of the (0001)-oriented QW structure are about fourfold smaller than those of the $(10\bar{1}0)$-oriented QW structure. This is mainly due to the increase in the spatial separation between the electron and hole wave functions in the conduction and valence bands. However, the matrix elements increase rapidly with the increasing crystal orientation because the spatial separation between the electron and hole wave functions decreases with the crystal orientation due to the reduced PZ and SP polarization effects.

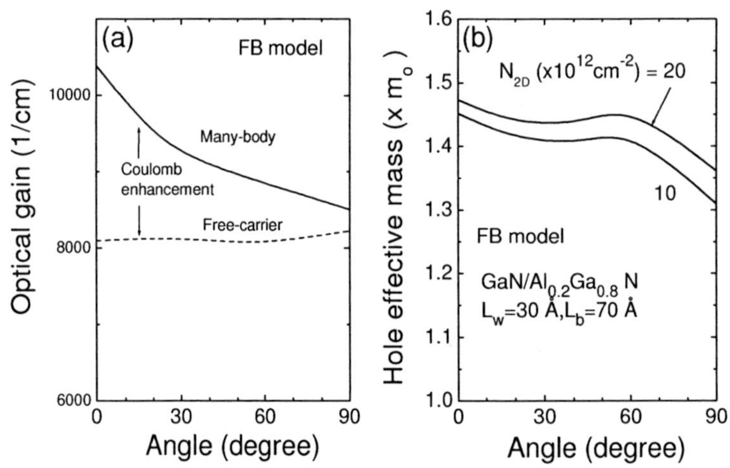


Figure 16: (a) $y'$-polarized TE optical gain and (b) the hole effective masses as functions of the crystal orientation of WZ GaN/$Al_{0.2}Ga_{0.8}N$ QWs ($L_w$=30 Å). Here, the FB model without spontaneous and piezoelectric polarization effects is used to study many-body effects. (From Ref. [26])

Figure 16 shows (a) $y'$-polarized TE optical gain and (b) the hole effective masses as functions of the crystal orientation of wurtzite GaN/$Al_{0.2}Ga_{0.8}N$ QWs ($L_w$=30 Å). Here, the FB model without spontaneous and piezoelectric polarization effects is used to study many-body effects. The difference between the solid (many-body) and dashed (free-carrier with bandgap renormalization) is the increase of the optical gain due to Coulomb enhancement contribution of the many-body effects. The Coulomb enhancement effect is the most dominant in a case of the (0001)-oriented QW ($\theta = 0^o$) and decreases with increasing crystal orientation. The decrease of Coulomb enhancement with the crystal angle can be explained by the decrease of the hole effective mass. To estimate the magnitude of the hole effective mass, we consider the parabolic band fitted to the lowest subband of the exact band structure. The effective mass is determined so that, for a given carrier density and the quasi-Fermi level for holes, the carrier density and the quasi-Fermi level agree with those of the exact band structure. Hole effective masses are averaged in the $k'_x$-$k'_y$ plane. It is shown that the effective mass gradually decreases with increasing crystal angle although they slightly increase near $\theta = 60^o$. The average hole effective mass in the non-(0001) QWs, particularly in the (10$\bar{1}$0) QW, is much smaller than that in the conventional (0001)-oriented QW. The larger effective mass leads to a larger exiton binding energy and, in turn, gives larger Coulomb enhancement. Thus, the Coulomb enhancement is gradually reduced with increasing crystal orientation. The crystal orientation dependence of the Coulomb enhancement for the SC model is also similar to that for the FB model because the crystal orientation dependence of the average hole effective mass for the SC model is similar to that for the FB model. These results suggest that the inclusion of the many-body effects is very important in order to give a guideline on device design. In the case of the free-carrier model with only bandgap renormalization, the optical gain increases with crystal orientation. This is because with increasing angle, the states constituting the topmost valence subband near the band edge change to predominantly $|Y'>$-like.

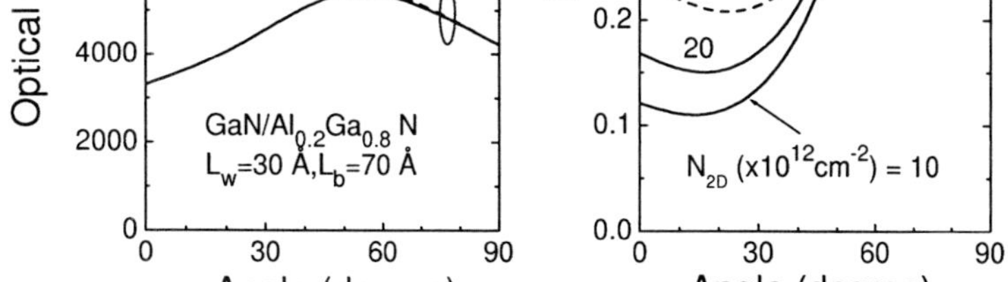


Figure 17: (a) Many-body optical gain and (b) normalized $y'$-polarized optical matrix elements at the band-edge ($k = 0$) as functions of the crystal orientation of WZ GaN/Al$_{0.2}$Ga$_{0.8}$N QWs ($L_w$=30 $\AA$) for the SC model. Optical matrix elements are averaged in the $k'_x$-$k'_y$ plane. (From Ref. [26])

Figure 17 shows (a) many-body optical gain and (b) normalized $y'$-polarized optical matrix elements at the band-edge ($k = 0$) as functions of the crystal orientation of wurtzite GaN/$Al_{0.2}Ga_{0.8}N$ QWs ($L_w$=30 Å) for the SC model. Optical matrix elements are averaged in the $k'_x$-$k'_y$ plane. For comparison, results for the FB model are also plotted. The optical gain and the optical matrix element are calculated at sheet carrier densities of $N_{2D} = 10$ and $20 \times 10^{12} cm^{-2}$. The FB model shows that the optical gain gradually decreases with increasing crystal angle, as discussed in Fig. 16. On the other hand, the SC model shows that the optical gain increases with increasing crystal angle and reaches a maximum near $\theta = 50^o$. These results can be analyzed by crystal orientation dependencies of the dipole moments. We find in Fig. 17(b) that the matrix element for the SC model is significantly reduced near the (0001) crystal orientation, compared to that of the FB model. This is because the spatial separation between the electron and hole wave functions in the conduction and valence bands increases due to larger internal fields in QW structures near the (0001) crystal orientation ($\theta = 0^o$). Thus, we know that the decrease of the optical gain near $\theta = 0^o$ is attributed to the decrease of the matrix elements because of the internal field due to spontaneous and piezoelectric polarization. However, the matrix elements increase rapidly with increasing crystal orientation. This is attributed to the fact that the internal field decreases with increasing crystal orientation. The reduction of the internal field results in the decrease of the spatial separation between the electron and hole wave functions and the increase of optical gain with increasing crystal angle.

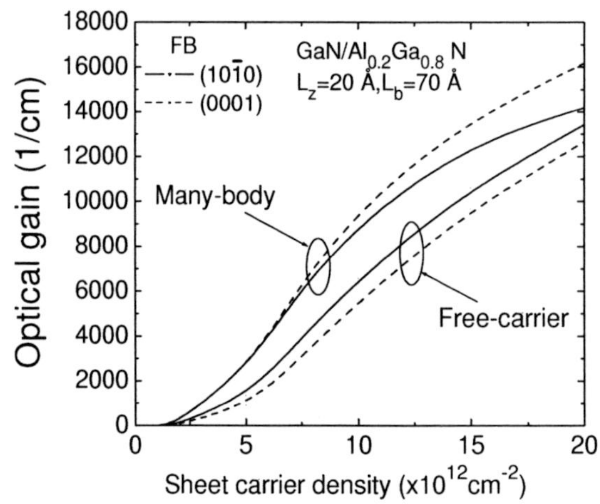


Figure 18: Optical gains as a function of carrier density for the free-carrier and many-body models of (0001)- and (10$\bar{1}$0)-oriented WZ GaN/$Al_{0.2}Ga_{0.8}N$ QWs ($L_w$=30 Å). Here, the FB model without the PZ and SP polarization is used for the (0001) orientation to investigate many-body effects. (From Ref. [27])

Figure 18 shows $y'$-polarized TE optical gains as a function of carrier density for the free-carrier and many-body models of (0001)- and (10$\bar{1}$0)-oriented WZ GaN/$Al_{0.2}Ga_{0.8}N$ QWs ($L_w$=30 Å). Here, the FB model without the PZ and SP polarization is used for the (0001) orientation to investigate many-body effects. In a case of the (10$\bar{1}$0) orientation, we always use the FB model since the PZ and SP polarization is expected to be zero at this orientation. The free-carrier model shows that the optical gain of the $(10\bar{1}0)$-oriented QW is enhanced over that of the (0001)-oriented QW. This is because, in the case of the $(10\bar{1}0)$-oriented QW structure, the states constituting the topmost valence subband near the band edge are predominantly $|Y' >$-like. This results in significantly enhanced $y'$-polarized optical matrix elements for the C1-$Y'1$ transition in the $(10\bar{1}0)$-oriented QW structure, in comparison with the $y'$-polarized optical matrix elements in the (0001)-oriented QW structure where $|X' >$ and $|Y' >$ are equally mixed. In addition, the average hole effective mass of the $(10\bar{1}0)$ crystal orientation is smaller than that of the (0001) crystal orientation However, considering the many-body effects, we know that the optical gain of the $(10\bar{1}0)$-oriented QW is smaller than that of the (0001)-oriented QW. That is, the Coulomb enhancement contribution to the optical gain of the (0001)-oriented QW is larger than that for the $(10\bar{1}0)$-oriented QW. This is because the average hole effective mass of the (0001)-oriented QW is larger than that of the $(10\bar{1}0)$-oriented QW. These results suggest that the inclusion of the many-body effects is very important in order to give a guideline on device design issues.

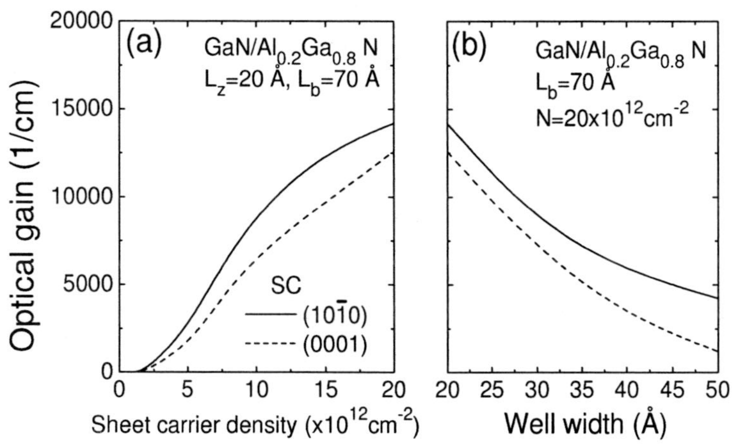


Figure 19: Many-body optical gain as a function of (a) carrier density and (b) well width of (0001)- and $(10\bar{1}0)$-oriented WZ GaN/$Al_{0.2}Ga_{0.8}N$ QWs ($L_b = 7\,nm$). (From Ref. [27])

In Fig. 19(a), we plot the many-body optical gain as a function of carrier density of (0001)- and $(10\bar{1}0)$-oriented WZ GaN/$Al_{0.2}Ga_{0.8}N$ QWs ($L_w = 3\,nm$). The optical gain of the (0001)-oriented QW is calculated self-consistently with the inclusion of the PZ and SP polarization. It is observed that the optical gain of the (0001)-oriented QW is significantly smaller than that of the $(10\bar{1}0)$-oriented QW. This is caused mainly by the decrease of the optical-matrix element due to the spatial separation of the electron and hole wave functions in the (0001)-oriented QW. Fig. 19(b) also shows the many-body optical gain as a function of the well width of (0001)- and $(10\bar{1}0)$-oriented WZ GaN/$Al_{0.2}Ga_{0.8}N$ QWs. The many-body optical gain of the $(10\bar{1}0)$-oriented QW is larger than that of the (0001)-oriented QW in a range of investigated well width. Also, the optical gain gradually increases with decreasing the well width for both orientations. The decrease of the well width leads to the increased subband energy spacing. This results in a more rapid increase in the quasi-Fermi levels for carriers. In a case of the (0001) orientation, the decrease of the optical gain is also caused by the decrease of the optical-matrix element due to the spatial separation of the electron and hole wave functions with increasing the well width.

### 3.2.2 InGaN/GaN QW Structures[24, 57, 58]

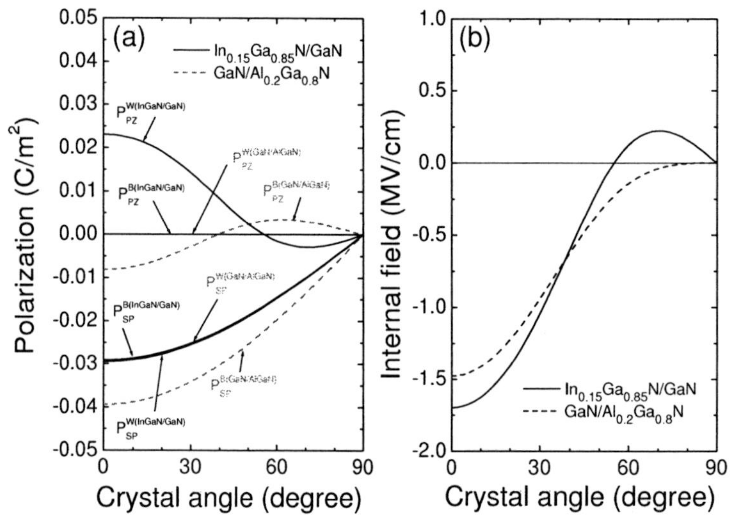


Figure 20: (a) Strain-induced PZ and SP polarization normal with respect to the growth plane for the well and the barrier and (b) the internal field as a function of the angle $\theta$ between the growth direction and the c-axis of the $In_{0.15}Ga_{0.85}N/GaN$ QW structure. For comparison, results (dashed line) of $GaN/Al_{0.2}Ga_{0.8}N$ QW structure are also plotted. (From Ref. [24])

Figure 20 shows (a) the strain-induced PZ and SP polarization normal with respect to the growth plane for the well and the barrier and (b) the internal field as a function of the angle $\theta$ between the growth direction and the c-axis of the $In_{0.15}Ga_{0.85}N/GaN$ QW structure. For comparison, results (dashed line) of $GaN/Al_{0.2}Ga_{0.8}N$ QW structure are also plotted. The GaN/AlGaN QW structure has the tensile strained AlGaN barrier and the unstrained GaN well. The normal PZ polarization is important because the QW structures have quantized energy levels along the growth direction. The normal PZ polarization leads to the accumulation or depletion of carriers at the interfaces and creates a piezoelectric field.[55] The GaN/AlGaN QW structure shows that there exists a relatively large difference of the SP polarization between the well and the barrier. On the other hand, the InGaN/GaN QW structure shows that the difference of the SP polarization between the well and the barrier is very small. Thus, in the case of the InGaN/GaN QW structure with a low In composition, neglecting SP polarization will be a good approximation. Note that the strain-induced normal PZ polarization $P_{PZ}$ is zero for the barrier in the InGaN/GaN structure and the well in the GaN/AlGaN structure. In the case of the GaN/AlGaN QW structure, it is shown that the internal field gradually decreases with increasing crystal angle. On the other hand, the InGaN/GaN QW structure shows that the internal field becomes zero near the crystal angle of $55^o$. This is because the sum of the PZ and SP polarization in the barrier is equal to that in the well. Also, it is observed that the internal field changes its sign when the crystal angle further increases. The internal field for both QW structures becomes zero for the $(10\bar{1}0)$ crystal orientation ($\theta = 90^o$) because the SP and strain-induced PZ polarization is zero.

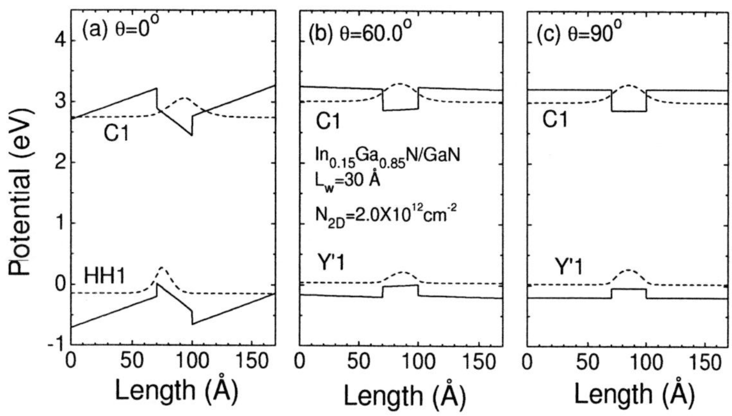


Figure 21: Potential profiles and the wave functions (C1, HH1, and Y'1) at zone center for the SC model with the SP and PZ polarization of WZ $In_{0.15}Ga_{0.85}N/GaN$ QW structures ($L_w = 3\ nm$) with crystal orientations of (a) $\theta = 0^o$, (b) $\theta = 60^o$, and (c) $\theta = 90^o$. (From Ref. [24])

Figure 21 shows the potential profiles and the wave functions (C1, HH1, and Y'1) at zone center for the SC model with the PZ and SP polarization of wurtzite $In_{0.15}Ga_{0.85}N/GaN$ QW structures ($L_w = 30\ \mathring{A}$) with crystal orientations of (a) $\theta = 0^o$, (b) $\theta = 60^o$, and (c) $\theta = 90^o$. The crystal orientations of $\theta = 0^o$ and $90^o$ correspond to (0001) and $(10\bar{1}0)$ orientation, respectively. Here, the barrier width $L_b$ is set to be 70 $\mathring{A}$. The SC solutions are obtained at the sheet carrier density of $N_{2D} = 2 \times 10^{12}\ cm^{-2}$. In the case of a (0001)-oriented InGaN/GaN QW, potential profiles show that there exist a large internal field in the well due to the difference of the PZ and SP polarization between the well and the barrier. The estimated field in the well is -1.69 MV/cm. The bottom of the potential well in the conduction band of a (0001)-oriented QW exits on the right side in the well. On the other hand, in the case of $\theta = 60^o$, it exists on the left side in the well because of the sign change of the internal field, as shown in Fig. 1(b). The SC model shows asymmetric electron and hole wavefunctions with respect to the well center and the spatial separation between these wavefunctions. In particular, a (0001)-oriented QW structure shows large spatial separation of the electron and hole wave functions. Also, the ground-state transition energy C1-HH1 is redshifted compared to the FB model due to the quantum-confined Stark effect. On the other hand, it is observed that the spatial separation of wavefunctions is significantly reduced with the increasing crystal orientation. This is due to the decrease of the PZ and SP polarization, which is given as a function of crystal orientation, as shown in Fig. 20(a).

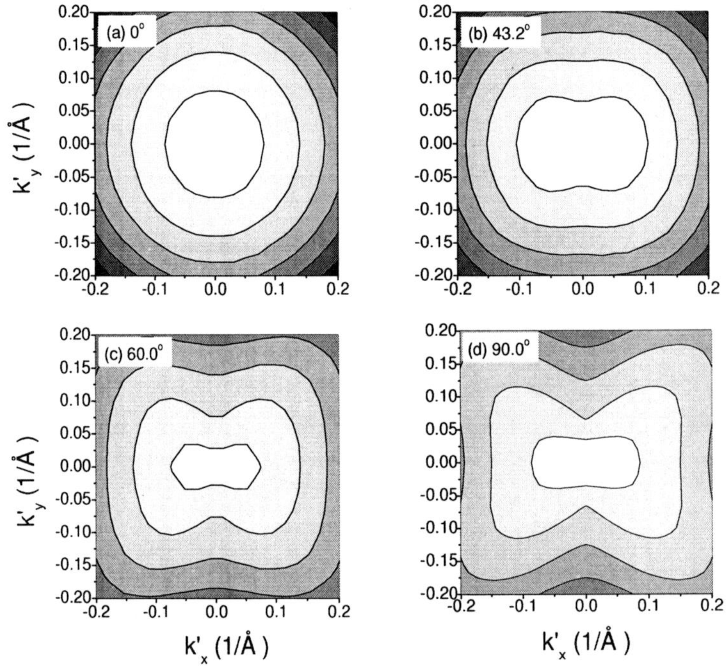


Figure 22: Constant energy contour in the in-plane wave vector space of WZ $In_{0.15}Ga_{0.85}N$/GaN QWs ($L_w$ = 30 Å) with angles of (a) $\theta = 0^o$, (b) $\theta = 43.2^o$, (c) $\theta = 60^o$, and (d) $\theta = 90^o$. Here, $k'_x$ and $k'_y$ are the in-plane wave vectors in the $(x', y', z')$ coordinate system. (From Ref. [58])

Figure 22 shows the constant energy contour in the in-plane wave vector space of WZ $In_{0.15}Ga_{0.85}N$/GaN QWs ($L_w = 30$ Å) with angles (a) $\theta = 0^o$, (b) $\theta = 43.2^o$, (c) $\theta = 60^o$, and (d) $\theta = 90^o$. Here, $k'_x$ and $k'_y$ are the in-plane wave vectors in the $(x', y', z')$ coordinate system. The angles $\theta = 0^o$ and $\theta = 90^o$ correspond to (0001) and $(10\bar{1}0)$ crystal orientations, respectively. The constant energy contour of QW structures with general crystal orientations shows anisotropy in the QW plane and has a twofold symmetry. On the other hand, the constant energy contour of the (0001) orientation is isotropic. It is observed that the average hole effective mass significantly decreases with increasing crystal angle. In particular, we know that the hole effective mass along ${k'}_y$ is smaller than that along ${k'}_x$. This is mainly attributed to the fact that the valence band effective mass parameter along ${k'}_y$ is smaller than that along ${k'}_x$, in addition to the increase in the energy difference between the first and second subbands with increasing crystal angle.[24]

Figures 23 shows the interband transition energies near the band edges as functions of the crystal angle $\theta$ between the growth direction and the c-axis of wurtzite $In_{0.15}Ga_{0.85}N$/GaN

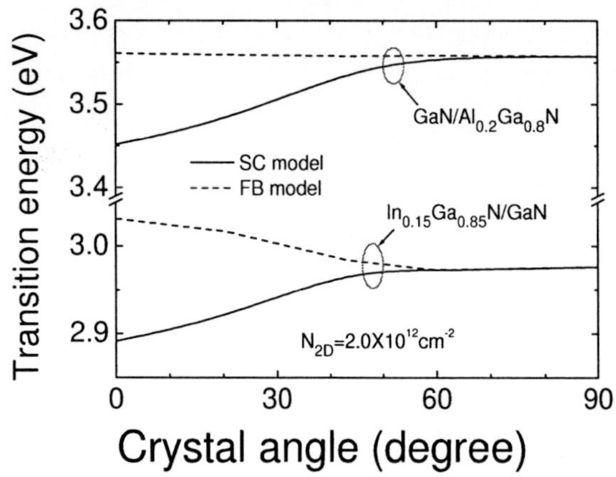


Figure 23: Interband transition energies near the band edges as functions of the crystal angle $\theta$ between the growth direction and the c-axis of WZ $In_{0.15}Ga_{0.85}N$/GaN and GaN/$Al_{0.2}Ga_{0.8}N$ QW structures ($L_w = 3\,nm$). (From Ref. [24])

and GaN/$Al_{0.2}Ga_{0.8}N$ QW structures ($L_w = 3\,nm$). For comparison, results (dashed line) for the FB model without considering the PZ and SP polarization effects are also plotted. The interband transition energy corresponds to the energy difference between the first conduction subband (C1) and the first valence subband. The transition energy of GaN/AlGaN QW is larger than that of InGaN/GaN because of its larger bandgap energy. The interband transition energies for the FB model slowly decrease with increasing crystal angle. On the other hand, the interband transition energy for the SC model gradually increases with the crystal angle due to the reduced PZ and SP polarization effects. It is shown that the interband transition energy for the SC model is significantly redshifted for crystal angles near $\theta$=0$^o$. These results suggest that an understanding of the role of the PZ and SP polarization as a function of crystal orientation is important to obtain guidelines for sample growth and device design issues.

Figure 24 shows the valence band structures for the SC model with the PZ and SP polarization of wurtzite $In_{0.15}Ga_{0.85}N$/GaN QW structures ($L_w = 3\,nm$) with crystal orientations of (a) $\theta = 0^o$, (b) $\theta = 20.6^o$, (c) $\theta = 60^o$, and (d) $\theta = 90^o$. Here, the naming of subbands for structures with the general crystal orientation follows the dominant composition of the wave function at the $\Gamma$ point in terms of $|X'>$, $|Y'>$, and $|Z'>$ bases. The SC solutions are obtained at the sheet carrier density of $N_{2D} = 2 \times 10^{12}\ cm^{-2}$. The result for (0001)-oriented QW structure ($\theta = 0^o$) shows a larger energy spacing between the first two subbands (HH1 and LH1) and higher subbands (HH2, LH2, and etc) due to the larger internal field. The increase of the subband energy spacing reduces the carrier population in the higher subbands. However, this effect is compensated by the decrease of the optical matrix element due to the spatial separation of the wave functions. The energy spacing between the first two subbands and higher subbands decreases with increasing crystal angle due to the reduction of the internal field. Also, it is observed that the effective masses of the first

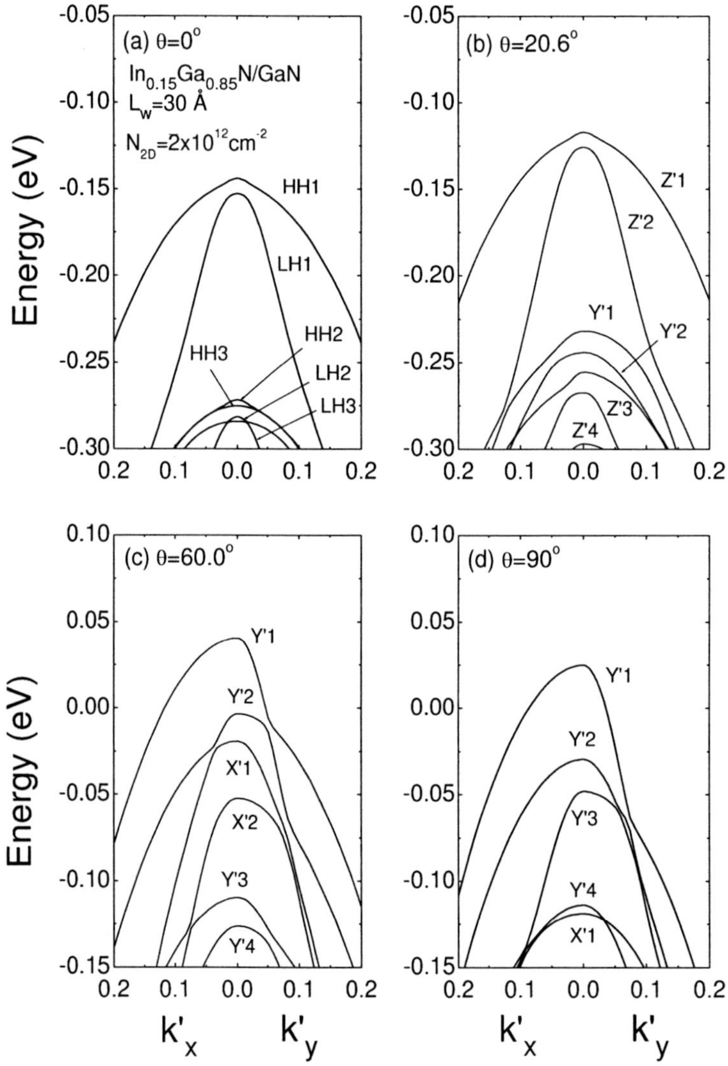


Figure 24: Valence band structures for the SC model with the SP and PZ polarization of WZ $In_{0.15}Ga_{0.85}N/GaN$ QW structures ($L_w = 3\ nm$) with crystal orientations of (a) $\theta = 0^o$, (b) $\theta = 20.6^o$, (c) $\theta = 60^o$, and (d) $\theta = 90^o$. (From Ref. [24])

subbands gradually decrease with increasing crystal angle. In particular, the effective mass along $k'_y$ around the topmost valence band is significantly reduced compared to that of the (0001)-oriented structure.

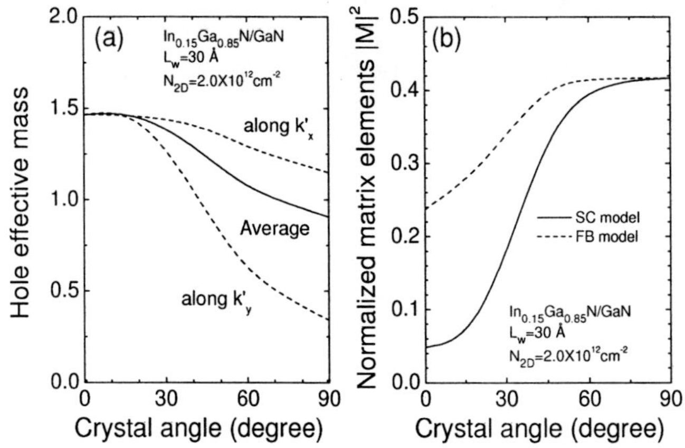


Figure 25: (a) Hole effective masses and (b) normalized $y'$-polarized optical matrix elements at band-edges as functions of crystal orientation of WZ $In_{0.15}Ga_{0.85}N/GaN$ QW structures ($L_w = 3\ nm$) with the SP and PZ polarization. (From Ref. [24])

Figure 25 shows (a) hole effective masses and (b) normalized $y'$-polarized optical matrix elements at band-edges as functions of crystal orientation of wurtzite $In_{0.15}Ga_{0.85}N/GaN$ QW structures ($L_w = 3\ nm$) with the PZ and SP polarization. The hole effective mass and optical matrix elements of the SC model are calculated at a sheet carrier density of $N_{2D} = 2 \times 10^{12} cm^{-2}$. To estimate the magnitude of the hole effective mass, we consider the parabolic band fitted to the lowest subband of the exact band structure. The effective mass is determined so that, for a given carrier density and the quasi-Fermi level for holes, the carrier density and the quasi-Fermi level agree with those of the exact band structure. Hence, the effective mass of the fitted parabolic band reflects an averaged density of states.[56] Hole effective masses and optical matrix elements are averaged in the $k'_x$-$k'_y$ plane because they show anisotropy in the QW plane for general crystal orientation. It is observed that the average hole effective mass is significantly reduced with increasing crystal angle. For QW structures with larger crystal angle, we know that the hole effective mass along $k'_y$ is smaller than that along $k'_x$. This is mainly attributed to the fact that the valence band effective mass parameter along $k'_y$ is smaller than that along $k'_x$.

The FB model shows that the $y'$-polarized optical matrix elements significantly depend on the crystal orientation. In particular, matrix elements of the $(10\bar{1}0)$-oriented QW structure are about seven times bigger than those of the (0001)-oriented QW structure. For QW structures with larger crystal angle, the optical matrix elements are dominated by the $y'$-polarization because the states constituting the topmost valence subband near the band edge are predominantly $|Y' >$-like. This results in significantly enhanced $y'$-polarized optical matrix elements for the C1-$Y'$1 transition in the $(10\bar{1}0)$-oriented QW structure in comparison with the $y$-polarized optical matrix elements in the (0001)-oriented QW structure where $|X>$ and $|Y>$ are equally mixed. In addition, the SC model shows that the matrix elements near the (0001) crystal orientation can be largely reduced compared to the FB model. However, the difference of matrix elements between the FB and SC models gradually decreases with increasing crystal angle because the spatial separation between the electron and hole wave functions decreases with the crystal orientation due to the reduced PZ and SP polarization effects. On the basis of these results, we know that the QW structure with a general crystal orientation show improved characteristics compared to the (0001)-oriented QW structure.

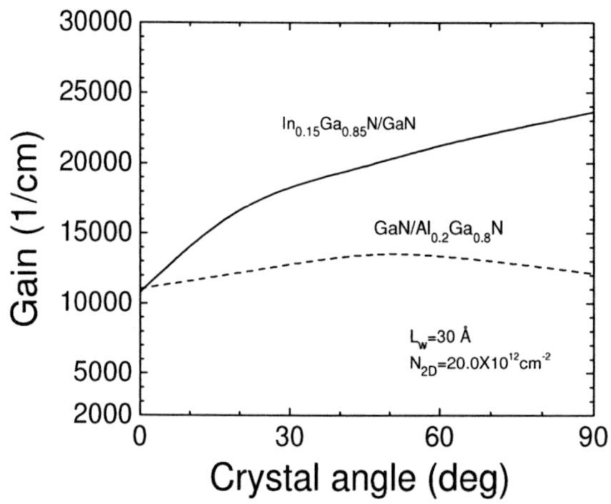


Figure 26: Many-body optical gain as a function of the crystal orientation of WZ $In_{0.15}Ga_{0.85}N$/GaN QWs ($L_w = 30$ Å) for the SC model. For comparison, results (dashed line) of GaN/$Al_{0.2}Ga_{0.8}N$ QW structure are also plotted. The optical gain is calculated at a sheet carrier density of $N_{2D}$=20 $\times$ $10^{12}$ cm$^{-2}$. (From Ref. [58])

Figure 26 shows many-body optical gain as a function of the crystal orientation of wurtzite $In_{0.15}Ga_{0.85}N$/GaN QWs ($L_w = 30$ Å) for the SC model. For comparison, results (dashed line) of the GaN/$Al_{0.2}Ga_{0.8}N$ QW structure are also plotted. The optical gain is calculated at a sheet carrier density of $N_{2D}$=20 $\times 10^{12} cm^{-2}$. It is found that the optical gain of the wurtzite InGaN/GaN QW structure largely increases with increasing crystal angle. On the other hand, in the case of the GaN/AlGaN QW structure, the improvement of the optical gain is relatively small compared to that of the wurtzite InGaN/GaN QW structure. The GaN/AlGaN QW structure shows that, with increasing crystal angle, the optical gain gradually increases and reaches a maximum near $\theta = 50^o$. Also, the optical gain decreases when the crystal angle further increases. For the QW structure with the (0001) crystal orientation, the difference in the optical gain between the InGaN/GaN and GaN/AlGaN QW structures is very small. However, in the case of that with the $(10\bar{1}0)$ crystal orientation, the optical gain of the InGaN/GaN QW structure is significantly larger than that of the GaN/AlGaN QW structure. We analyze these results by considering characterization parameters such as the averaged effective mass, squared optical matric element, and quasi-Fermi energy separation.

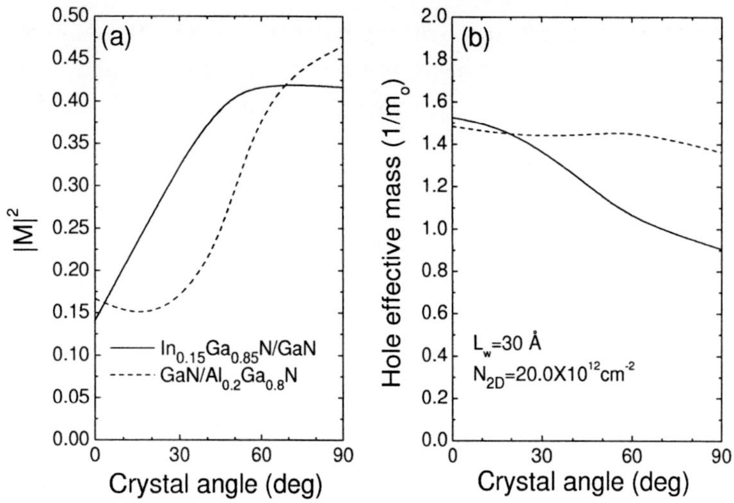


Figure 27: (a) Normalized $y'$-polarized optical matrix elements and (b) average hole effective masses as functions of crystal orientation of WZ $In_{0.15}Ga_{0.85}N/GaN$ QWs ($L_w$ = 30 Å). For comparison, results (dashed line) of $GaN/Al_{0.2}Ga_{0.8}N$ QW structure are also plotted. (From Ref. [58])

Figure 27 shows the (a) normalized $y'$-polarized optical matrix elements and (b) average hole effective masses as functions of the crystal orientation of wurtzite $In_{0.15}Ga_{0.85}N/GaN$ QWs ($L_w$ = 30 Å). For comparison, results (dashed line) of the $GaN/Al_{0.2}Ga_{0.8}N$ QW structure are also plotted. The optical matrix elements and the hole effective mass of the SC model are calculated at a sheet carrier density of $N_{2D} = 20 \times 10^{12} cm^{-2}$. To estimate the magnitude of the hole effective mass, we consider a parabolic band fitted to the lowest subband of the exact band structure. The effective mass is determined so that, for a given carrier density and the quasi-Fermi level for holes, the carrier density and the quasi-Fermi level agree with those of the exact band structure. Hence, the effective mass of the fitted parabolic band reflects an averaged density of states.[56] The optical matrix elements and the hole effective masses are averaged in the $k'_x$-$k'_y$ plane.

The InGaN/GaN QW structure shows that the matrix element rapidly increases with crystal angle and saturates near $\theta = 60^o$. In the case of the GaN/AlGaN QW structure, the matrix element slightly decreases near $\theta = 20^o$ and then rapidly increases when the crystal angle exceeds $\theta = 30^o$. The increase of the matrix element is mainly attributed to the fact that the internal field decreases with increasing crystal orientation. The reduction of the internal field results in the decrease in the spatial separation between the electron and hole wave functions and the increase in the optical matrix element with increasing crystal angle. The matrix element near $\theta = 90^o$ is more than twofold larger than that near $\theta = 0^o$. The effective mass of the InGaN/GaN QW structure is shown to largely decrease with increasing crystal angle. In particular, the hole effective mass of $(10\bar{1}0)$ InGaN/GaN QW ($\theta = 90^o$) is significantly reduced compared to that of the conventional (0001)-oriented

QW. For example, the hole effective masses for the (0001) and $(10\bar{1}0)$ QWs are $1.53m_o$ and $0.91m_o$, respectively. On the other hand, in the case of the GaN/AlGaN QW structure, the reduction effect of the hole effective mass is relatively small compared to that of the InGaN/GaN QW structure.

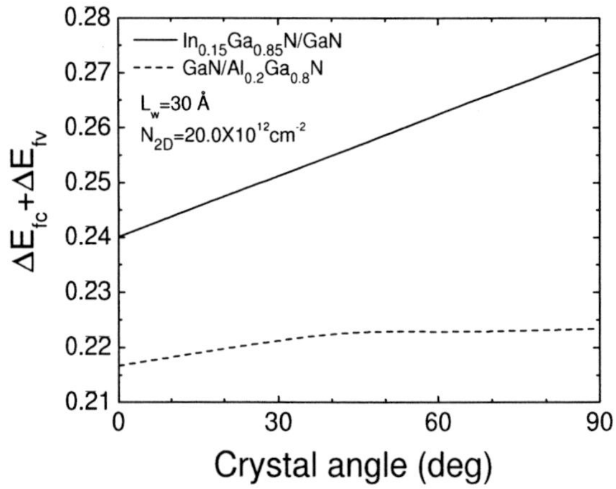


Figure 28: Quasi-Fermi level separation $\Delta E_{fc}$+ $\Delta E_{fv}$ as a function of the crystal orientation of WZ $In_{0.15}Ga_{0.85}N$/GaN QWs ($L_w = 30$ Å). For comparison, results (dashed line) of GaN/$Al_{0.2}Ga_{0.8}N$ QW structure are also plotted. (From Ref. [58])

In Fig. 28, we plot the quasi-Fermi level separation $\Delta E_{fc}$+$\Delta E_{fv}$ as a function of the crystal orientation of wurtzite $In_{0.15}Ga_{0.85}N$/GaN QWs ($L_w = 30$ Å). For comparison, the results (dashed line) of GaN/$Al_{0.2}Ga_{0.8}N$ QW structure are also plotted. The quasi-Fermi level separation $\Delta E_{fc}$ ($\Delta E_{fv}$) is defined as the energy difference between the quasi-Fermi level and the ground-state energy in the conduction band (the valence band). The quasi-Fermi level separation of the GaN/AlGaN QW structure slightly increases with increasing crystal angle. On the other hand, the InGaN/GaN QW structure shows a rapid increase in the quasi-Fermi level separation compared to the GaN/AlGaN QW structure. Also, the quasi-Fermi level separation of the InGaN/GaN QW structure is shown to be larger than that of the GaN/AlGaN QW structure. This is due to the reduction of the hole effective mass shown in Fig. 27(b). Thus, in the case of the InGaN/GaN QW structure, the increase in the quasi-Fermi level separation is responsible for the rapid increase in the optical gain since both systems show similar dependences of the matrix element on crystal angle.

Figure 29 show (a) interband transition energies as functions of the carrier density of (0001)- and $(10\bar{1}0)$-oriented WZ $In_{0.15}Ga_{0.85}N$/GaN QW structures ($L_w = 30$ Å). The interband transition energy corresponds to the energy difference between the first conduction subband (C1) and the first valence subband. The ground states for holes in the valence band are HH and $Y'1$ bands for the (0001)- and $(10\bar{1}0)$-oriented QW structures, respectively. The interband transition energy for the (0001)-oriented QW structure is significantly redshifted at low carrier densities compared to that for the $(10\bar{1}0)$-oriented QW structure. However, the interband transition energy rapidly increases with increasing carrier density due to the

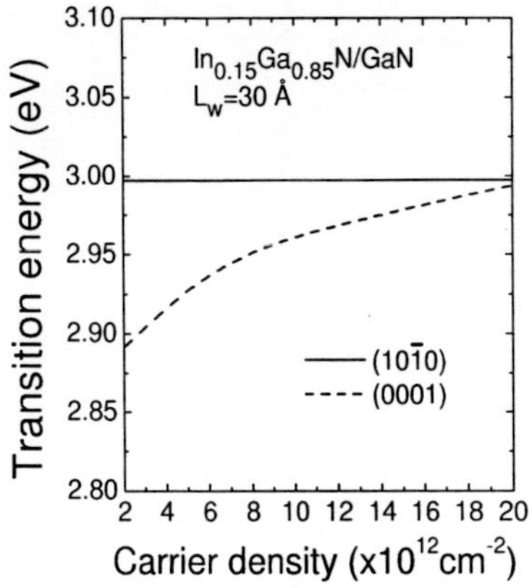


Figure 29: (a) Interband transition energies as functions of the carrier density for the many-body model of (0001)- and (10$\bar{1}$0)-oriented WZ $In_{0.15}Ga_{0.85}N$/GaN QW structures ($L_w$ = 30 $\AA$). (From Ref. [57])

the increase of the screening electric field to compensate the SP and PZ polarization effects. In the case of carrier density of $20 \times 10^{12} cm^{-2}$, both QW structures have similar transition energies.

Figure 30 shows $y'$-polarized TE optical gain as a function of (a) carrier density and (b) radiative current density for the many-body model of (0001)- and (10$\bar{1}$0)-oriented WZ $In_{0.15}Ga_{0.85}N$/GaN QW structures ($L_w$ = 30 $\AA$). In the case of the (10$\bar{1}$0) orientation, we use the FB model since the PZ and SP polarization is expected to be zero at this orientation[53]. The optical gain of the (10$\bar{1}$0)-oriented QW is largely enhanced over that of the (0001)-oriented QW in the investigated range of the carrier density. This is because, in the case of the (10$\bar{1}$0)-oriented QW structure, the states constituting the topmost valence subband near the band edge are predominantly $|Y' >$-like and the internal field due to the PZ and SP polarization is negligible. This results in significantly enhanced $y'$-polarized optical matrix elements in the (10$\bar{1}$0)-oriented QW structure, in comparison with the $y'$-polarized optical matrix elements in the (0001)-oriented QW structure where $|X' >$ and $|Y' >$ are equally mixed. Assuming a threshold current density $J_{th}$ of about 4.3 kA/$cm^2$[46], the threshold optical gain is estimated to be about 13000 $cm^{-1}$ for the (0001)-oriented QW structure. If we take this value as a threshold optical gain for the (10$\bar{1}$0)-oriented QW structure, we can obtain the threshold current density $J_{th}$ of about 2.4 kA/$cm^2$, which is about half that of the (0001)-oriented QW structure. Thus, we expect that improved lasing characteristics are obtained by using the (10$\bar{1}$0) crystal orientation. The low gain for (10$\bar{1}$0)-oriented structures at low radiative current values is attributed to the fact that the radiative recombination coefficent of (10$\bar{1}$0)-oriented structures is much larger than that of (0001)-oriented structures at low radiative current values, as discussed below.

Figure 31 shows the radiative recombination coefficient $B_{eff}$ as a function of the carrier density for (0001)- and (10$\bar{1}$0)-oriented WZ $In_{0.15}Ga_{0.85}N$/GaN QW structures ($L_w$ = 30 $\AA$). The radiative recombination coefficient is obtained from the calculated $J_{rad}$ by

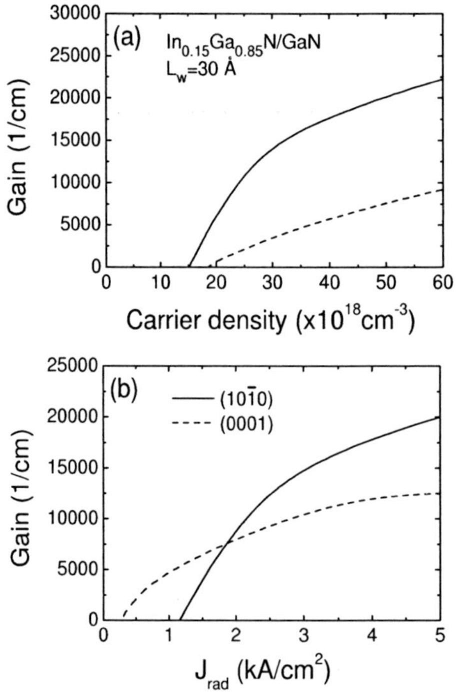


Figure 30: $y'$-polarized TE optical gain as a function of (a) carrier density and (b) radiative current density for the many-body model of (0001)- and (10$\bar{1}$0)-oriented WZ $In_{0.15}Ga_{0.85}N$/GaN QW structures ($L_w = 30\ \AA$). (From Ref. [57])

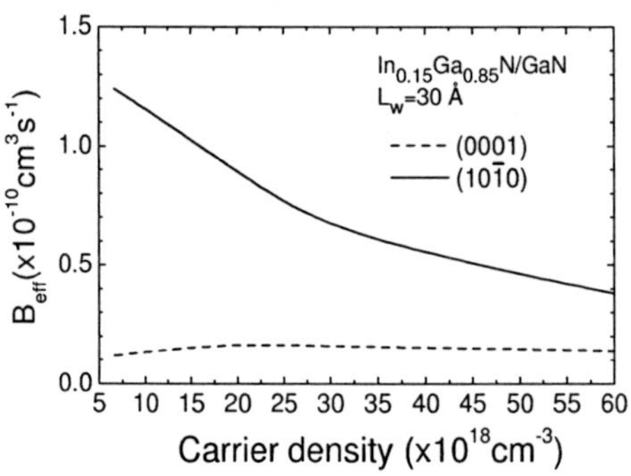


Figure 31: Radiative recombination coefficient $B_{eff}$ as a function of the carrier density for (0001)- and (10$\bar{1}$0)-oriented WZ $In_{0.15}Ga_{0.85}N$/GaN QW structures ($L_w = 30\ \AA$). The radiative recombination coefficient is obtained from the calculated $J_{rad}$ by $B_{eff} = J_{rad}/(eL_wN^2)$. (From Ref. [57])

$B_{eff} = J_{rad}/(eL_wN^2)$. The $B_{eff}$ of the (10$\bar{1}$0)-oriented QW structure rapidly decreases with increasing carrier density. On the other hand, in the case of the (0001)-oriented QW structure, the $B_{eff}$ is nearly independent of the carrier density. It is observed that $B_{eff}$ of the (10$\bar{1}$0)-oriented QW structure is much larger than the (0001)-oriented QW structure in the range of the lower carrier density. This results in significantly enhanced optical gain in the (10$\bar{1}$0)-oriented QW structure, in comparison with that in the (0001)-oriented QW structure, as shown in Fig. 30(a). The smaller $B_{eff}$ observed in the case of the (0001)-oriented QW structure can be explained by the fact that the internal field due to the PZ and SP polarization is dominant in the well.

# 4 Summary

Crystal orientation effects on electronic and optical properties of strained GaN semiconductors are reviewed. Analytical expressions for the band gap, wave function, and momentum matrix element of bulk GaN semiconductors with an arbitrary crystal orientation are shown using a block-diagonalized Hamiltonian. Self-consistent (SC) solutions considering the spontaneous (SP) and piezoelectric (PZ) polarizations for valence-band structures of the QW structure with an arbitrary crystal orientation are obtained by solving the Schrödinger equation for electrons, the $6 \times 6$ Hamiltonian for holes, and Poisson's equation iteratively. The optical gain is investigated using the non-Markovian gain model with many-body effects. The optical gain increases with increasing crystal angle, which is mainly due to the increase of the matrix elements with increasing crystal angle. In particular, the optical gain of the (10$\bar{1}$0)-oriented QW structure is largely enhanced over that of the (0001)-oriented QW structure. This can be explained by the fact that the optical matrix elements near the band edge ($k$=0) of the (10$\bar{1}$0)-oriented QW structure significantly increase compared to that of the (0001)-oriented structure, in addition to the reduction of the effective mass. The threshold current density $J_{th}$ of the (10$\bar{1}$0)-oriented QW structure is expected to be about half that of the (0001)-oriented QW structure.

# References

[1] S. Nakamura, M. Senoh, S. Nagahama, N. Iwasa, T. Matsushita and T. Mukai: *Proc. 2nd Int. Symp. Blue Laser and Light Emitting Diodes*, 371 (1998).

[2] S. L. Chuang, *Physics of Optoelectronic Devices*, Chapter 4 (New York: Wiley, 1995).

[3] W. Fang and S. L. Chuang, *Appl. Phys. Lett.* **67**, 751(1995).

[4] M. Suzuki and T. Uenoyama, Japan. *J. Appl. Phys.* **35**, L953(1996).

[5] S. H. Park and S. L. Chuang, *Appl. Phys. Lett.* **72**, 287(1998).

[6] S. L. Chuang and C .S. Chang, *Phys. Rev. B* **54**, 2491 (1996).

[7] M. Kumagai, S. L. Chuang, and H. Ando, *Phys. Rev. B* **57**, 15303 (1998).

[8] G. Martin, A. Botchkarev, A. Rockett, and H. Morkoç, *Appl. Phys. Lett.* **68**, 2541 (1996).

[9] A. D. Bykhovski, B. L. Gelmont, and M. S. Shur, *J. Appl. Phys.* **81**, 6332 (1997).

[10] F. Bernardini, V. Fiorentini, and D. Vanderbilt, *Phys. Rev.B* **56**, 10024 (1997).

[11] O. Ambacher, J. Smart, J. R. Shealy, N. G. Weimann, K. Chu, M. Murphy, W. J. Schaff, L. F. Eastman, R. Dimitrov, L. Wittmer, M. Stutzmann, W. Rieger, and J. Hilsenbeck, J. Appl. Phys. **85**, 3222 (1999).

[12] R. Langer, J. Simson, V. Oritz, N. T. Pelekanos, A. Barski, R. André, M. Godlewski, *Appl. Phys. Lett.* **74**, 3827 (1999).

[13] M. Leroux, N. Grandjean, J. Massies, B. Gil, P. Lefebvre, and P. Bigenwald, *Phys. Rev.B* **60**, 1496 (1999).

[14] J. Wang, J. B. Jeon, Yu. M. Sirenko, and K. W. Kim, *IEEE Photon. Technol. Lett.***9**, 728(1997).

[15] S. H. Park and S. L. Chuang, *Appl. Phys. Lett.* **72**, 3103(1998).

[16] S. H. Park and S. L. Chuang, *J. Appl. Phys.* **87**, 353 (2000).

[17] S. H. Park and S. L. Chuang, *Appl. Phys. Lett.* **72**, 1981(2000).

[18] K. Domen, K. Horino, A. Kuramata, and T. Tanahashi, *Appl. Phys. Lett.* **70**, 987 (1997).

[19] A. Niwa, T. Ohtoshi, and T. Kuroda, *Appl. Phys. Lett.* **70**, 2159 (1997).

[20] Y. C. Yeo, T. C. Chong, and M. F. Li, *IEEE J. Quantum Electron.* **34**, 1270 (1998).

[21] T. Ohtoshi, A. Niwa, and T. Kuroda, *J. Appl. Phys.* **82**, 1518 (1997).

[22] T. Takeuchi, H. Amano and I. Akasaki, *Jpn. J. Appl. Phys.* **39**, 413 (2000).

[23] S. H. Park, *Jpn. J. Appl. Phys.* **39**, 3478 (2000).

[24] S. H. Park, *J. Appl. Phys.* **91**, 9904 (2002).

[25] S. H. Park, *J. Korean Phys. Soc.* **41**, 813 (2002).

[26] S. H. Park and S. L. Chuang, *Semicond. Sci. Technol.* **17**, 686 (2002).

[27] S. H. Park, *Appl. Phys. Lett.* **77**, 4095 (2000).

[28] S. H. Park, *J. Appl. Phys.* **93**, 9665 (2002).

[29] S. H. Park, *J. Korean Phys. Soc.* **42**, 344 (2003).

[30] W. W. Chow, A. Knorr, and S. W. Koch, *Appl. Phys. Lett.* **57**, 754 (1995).

[31] S. H. Park and S. L. Chuang, *Phys. Rev. B* **59**, 4725 (1999).

[32] J. M. Hinckley and J. Singh, *Phys. Rev. B* **42**, 3546 (1990).

[33] J. F. Nye, F. R. S., *Physical Properties of Crystals*, Chapters 5 and 6 (Clarendon·Oxford, 1989).

[34] D. L. Smith and C. Mailhiot, *J. Appl. Phys.***63**, 2717 (1988).

[35] G. Bastard, *Wave Mechanics Applied to Semiconductor Heterostructures*, Halsted Press, Wiley, New York, Chap.III, 1988.

[36] D. L. Smith and C. Mailhiot, *Rev. Mod. Phys.* **62**, 173 (1990).

[37] M. Abramowitz and I. A. Stegun, *Handbook of Mathematical Functions*, p. 17 (Dover, New York, 1972).

[38] A. Shikanai, T. Azuhata, T. Sota, S. Chichibu, A. Kuramata, K. Horino, and S. Nakamura, *J. Appl. Phys.***81**, 417(1997).

[39] S. L. Chuang and C .S. Chang, *Semicond. Sci. Technol.***12**, 252(1997).

[40] H. P. Maruska and J. J. Tietjen, *Appl. Phys. Lett.* **15**, 327(1969).

[41] W. M. Yim, E. J. Stofko, P. J. Zanzucchi, J. I. Pankove, M. Ettenberg, and S. L. Gilbert, *Appl. Phys. Lett.* **44**, 292(1973).

[42] K. Kim, W. R. L. Lambrecht, and B. Segall, *Phys. Rev. B* **53**, 16310 (1996).

[43] Physics of Group IV Elements and III-V Compounds, edited by K.H.Hellwege and O.Madelung, Landolt-Bo..rnstein, New Series, Group III, Vol.17, Pt.a (Springer-Verlag, Berlin, 1982).

[44] T. L. Tansley and C. P. Foley, *J. Appl. Phys.* **59**, 3241 (1986).

[45] M. Suzuki, T. Uenoyama, and A. Yanase, *Phys. Rev. B* **52**, 8132 (1995).

[46] S. Nakamura and G. Fasol, *The Blue Green Diode* (Springer, Berline, 1997), Chap.3.

[47] W. W. Chow, M. Hagerott, A. Gimdt, and S. W. Koch, *IEEE J. Select. Topics Quantum Electron.* **4**, 514 (1998).

[48] T. Ohtoshi, A. Niwa, and T. Kuroda, Japan. *J. Appl. Phys.***35**, 1566(1996).

[49] V. W. L. Chin, T. L. Tansley, and Osotchan, *J. Appl. Phys.***75**, 7365(1994).

[50] A. Polian, M. Grimsditch, and I. Grzegory, *J. Appl. Phys.***79**, 3343(1996).

[51] C. Y-P. Chao and S. L. Chuang, *Phys. Rev. B* **46**, 4110 (1992).

[52] D. Ahn, *Prog. Quantum Electron.* **21**, 249 (1997).

[53] S. H. Park, S. L. Chuang, and D Ahn, *Semicond. Sci. Technol.* **15**, 203 (2000).

[54] W. W. Chow, S. W. Koch, and M. Sergent III, *Semiconductor-Laser Physics*, (Springer, Berli, 1994).

[55] A. Bykhovski, B. Gelmont, and S. Shur, *Appl. Phys. Lett.* **63**, 2243(1993).

[56] S. Seki, T. Yamanaka, W. Lui and K. Yokoyama, *J. Appl. Phys.* **75**, 1299 (1994).

[57] S. H. Park, *J. Appl. Phys.* **93**, 9665 (2003).

[58] S. H. Park, *Jpn. J. Appl. Phys.* **42**, L170 (2003).

# INDEX

## L



## M



## N



## O



## P



## Q



## R

## S



## T



## U



## V



## W



## X